T0321793

Membrane Computing for Distributed Control of Robotic Swarms:

Emerging Research and Opportunities

Andrei George Florea
Politehnica University of Bucharest, Romania

Cătălin Buiu
Politehnica University of Bucharest, Romania

A volume in the Advances in
Computational Intelligence and
Robotics (ACIR) Book Series

DISSEMINATOR of KNOWLEDGE

www.igi-global.com

Published in the United States of America by
  IGI Global
  Information Science Reference (an imprint of IGI Global)
  701 E. Chocolate Avenue
  Hershey PA, USA 17033
  Tel: 717-533-8845
  Fax: 717-533-8661
  E-mail: cust@igi-global.com
  Web site: http://www.igi-global.com

Library of Congress Cataloging-in-Publication Data

Names: Florea, Andrei George, 1991- author. | Buiu, Catalin, 1968- author.
Title: Membrane computing for distributed control of robotic swarms :
  emerging research and opportunities / by Andrei George Florea and Catalin
  Buiu.
Description: Hershey, PA : Information Science Reference, [2017] | Includes
  bibliographical references.
Identifiers: LCCN 2016057756| ISBN 9781522522805 (hardcover) | ISBN
  9781522522812 (eISBN)
Subjects: LCSH: Robots--Control systems. | Natural computation. | Swarm
  intelligence. | Self-organizing systems. | Distributed parameter systems.
Classification: LCC TJ211.35 .F57 2017 | DDC 629.8/926384--dc23 LC record available at https://
lccn.loc.gov/2016057756

This book is published in the IGI Global book series Advances in Computational Intelligence and Robotics (ACIR) (ISSN: 2327-0411; eISSN: 2327-042X)

British Cataloguing in Publication Data
A Cataloguing in Publication record for this book is available from the British Library.

All work contributed to this book is new, previously-unpublished material.
The views expressed in this book are those of the authors, but not necessarily of the publisher.

For electronic access to this publication, please contact: eresources@igi-global.com.

Advances in Computational Intelligence and Robotics (ACIR) Book Series

ISSN:2327-0411
EISSN:2327-042X

Editor-in-Chief: Ivan Giannoccaro, University of Salento, Italy

MISSION

While intelligence is traditionally a term applied to humans and human cognition, technology has progressed in such a way to allow for the development of intelligent systems able to simulate many human traits. With this new era of simulated and artificial intelligence, much research is needed in order to continue to advance the field and also to evaluate the ethical and societal concerns of the existence of artificial life and machine learning.

The **Advances in Computational Intelligence and Robotics (ACIR) Book Series** encourages scholarly discourse on all topics pertaining to evolutionary computing, artificial life, computational intelligence, machine learning, and robotics. ACIR presents the latest research being conducted on diverse topics in intelligence technologies with the goal of advancing knowledge and applications in this rapidly evolving field.

COVERAGE

- Automated Reasoning
- Heuristics
- Agent technologies
- Brain Simulation
- Computational Logic
- Pattern Recognition
- Intelligent control
- Robotics
- Computational Intelligence
- Artificial Life

IGI Global is currently accepting manuscripts for publication within this series. To submit a proposal for a volume in this series, please contact our Acquisition Editors at Acquisitions@igi-global.com or visit: http://www.igi-global.com/publish/.

Titles in this Series

For a list of additional titles in this series, please visit:
http://www.igi-global.com/book-series/advances-computational-intelligence-robotics/73674

For an enitre list of titles in this series, please visit:
http://www.igi-global.com/book-series/advances-computational-intelligence-robotics/73674

DISSEMINATOR of KNOWLEDGE

www.igi-global.com

701 East Chocolate Avenue, Hershey, PA 17033, USA
Tel: 717-533-8845 x100 • Fax: 717-533-8661
E-Mail: cust@igi-global.com • www.igi-global.com

Table of Contents

Foreword

From Biology to Robotics

Membrane Computing (MC) is a branch of Natural Computing initiated, in the fall of 1998, in the aim of abstracting computer science ideas (starting from data structures, operations with them, ways to control these operations, and ending with computing models) from the architecture and the functioning of biological cells, considered alone or in communities (tissues, organs – brain included, colonies of bacteria, organisms). The domain developed so fast (at the theoretical level), that already in 2000 an international series of meetings was started (called Workshop on MC for the first ten editions, continued as a Conference on MC after that – with a companion Asian Conference on MC in the last years), while in February 2003 the Institute for Scientific Information, ISI, identified the initial paper as fast breaking and MC as emerging research area in computer science. In 2016, an International MC Society, IMCS, was founded – see its bi-annually Bulletin at http://membranecomputing. net/IMCSBulletin. Another comprehensive source of information is the MC web site, at http://ppage.psystems.eu.

It is worth pointing out that the initial motivation of MC and the initial developments were of a theoretical computer science type (power and efficiency of computing models, in comparison with "standard" models, such as Turing machines and formal grammars), but soon also applications have been reported. As expected, the first ones were in biology and biomedicine: the biologists need models of cells, MC is explicitly born looking to the cell, hence the proposed models (currently called P systems) look adequate and understandable (also, scalable, easily programmable, etc.) to the biologist.

Then, also expected for any model which is abstract enough and general enough, applications in other areas were reported: ecology, linguistics, economics, cryptography, computer graphics.

All these areas have several features in common: they deal with discrete data (in the mathematical sense), which makes inappropriate continuous mathematical models of the differential equations type; they deal with distributed, parallel processes; probabilities and other related features (such as the stoichiometry and reaction rates in biochemistry) play a central role; a large number of "agents" evolve together according to a large number of "reactions", with an emergent behavior which cannot be predicted analytically, hence computer simulations are necessary; scalability of models and programs is a crucial issue when processing real data... In short, the biochemical metaphor together with the (cell-like of tissue-like) membrane localization are powerful and ubiquitous paradigms.

A series of applications of these kinds – most of them in biology – were presented in:

- Gabriel Ciobanu, Gheorghe Păun, Mario J. Pérez-Jiménez, eds.: *Applications of Membrane Computing*, Springer-Verlag, Berlin, 2006.
- Pierluigi Frisco, Marian Gheorghe, Mario J. Pérez-Jiménez, eds.: *Applications of Membrane Computing in Systems and Synthetic Biology*, Springer-Verlag, Berlin, 2014, as well as in
- Gheorghe Păun, Grzegorz Rozenberg, Arto Salomaa, eds.: *The Oxford Handbook of Membrane Computing*, Oxford University Press, 2010.

However, in the last years, a series of applications of P systems were carried out in areas which are/look far from biology: approximate optimization, engineering/technology, and – also the case of the present work – robots control. Both "standard" MC models and "less biological" versions of P systems are used in these areas, such as numerical P systems, fuzzy neural P systems, various distributed evolutionary computing models with the distribution organized in a cell-like manner, etc.

Continuing the previous references, I mention here a book,

- Gexiang Zhang, Jixiang Cheng, Tao Wang, Xueyuan Wang, Jie Zhu: *Membrane Computing: Theory and Applications*, Science Press, Beijing, China, 2015 (unfortunately, available only in Chinese), as well as two PhD theses recently completed in the research group of the authors of the present book:
- A.B. Pavel: *Development of Robot Controllers Using Membrane Computing*. PhD Thesis, Department of Automatic Control and Systems Engineering, Politehnica University of Bucharest, Romania, July 2015.
- C.I. Vasile: *Distributed Control for Multi-Robot Systems*. PhD Thesis, Department of Automatic Control and Systems Engineering, Politehnica University of Bucharest, Romania, July 2015.

The present book completes in a beautiful way the landscape. It is focused on recent developments in applying MC tools to robots control, using especially P colonies (communities of very simple "cells" evolving in a common environment) in approaching a rather modern issue, modelling/controlling swarms of robots. From Membrane Computing theoretical background details necessary to the subsequent developments, in addressing swarm robotics, to software and examples of (experimental) applications, the text invites the reader not only to consider new tools in his/her robotics applications, but also to explore a modern research direction which is both promising and fast evolving.

Gheorghe Păun
Romanian Academy, Bucharest

Preface

BACKGROUND

The visible side of nature has always been a source of inspiration not only for poets, artists, sages, philosophers, but also for remarkable scientists and researchers during the history of mankind. Remember in this context, for example, the works of Leonardo da Vinci, e.g. his *Codex on the Flight of Birds* which can be considered as an early study on the nature of flight and as providing suggestions for designing flying machines.

The last decades have witnessed a tremendous progress in science and technology that allowed a better visualization and understanding of the profound biological and biochemical building blocks of life. Consequently, the invisible side of nature has become a source of knowledge, action, and inspiration as well. New powerful visualization techniques (such as cryo-electron tomography, or the nuclear magnetic resonance) allowed us to delve into the intricate mechanisms of life and understand how a cell works. Cell biologists have benefitted the most from this unprecedented flow of data which still waits to be deciphered and transformed into valuable information and actionable knowledge.

We now know that all cells store their hereditary information using the same linear chemical code of DNA. Due to technological advances of the sequencing machines we know the complete genome sequences for humans and many other species (but still only a minor part of the huge number of the estimated 10 million living species on Earth). We know that a mechanism of templated polymerization allows the cells to replicate their hereditary information. The DNA is made of simple subunits, called nucleotides and a DNA strand can be seen as a string over an alphabet with 4 letters (A, T, C, G). This hereditary information is transcribed into strings over the RNA alphabet (A, U, C, G) and then expressed into proteins, the primordial agents of the biochemical activities. With some minor exceptions, the rules governing this translation into proteins (the genetic code) are the same for every living form.

The cell can be considered as a very complex dynamic biochemical system, with thousands of chemical reactions taking place simultaneously. The vast majority of cell reactions can take place only in the presence of specialized proteins (enzymes). The proteins also play a variety of other functional and structural roles. Each and every cell is separated from the outer environment by the plasma membrane. This membrane acts as a barrier whose function is to maintain different concentrations of solutes in the cytosol as compared to the internal compartments and to the outer environment (extracellular fluid). There are specialized proteins that allow the passage of small molecules across the cell's membrane, proteins that are embedded in the membrane. These proteins are transporters (proteins with moving parts) and channels (forming hydrophilic pores). The importance of membranes as separators of chemical reactions is evident in the presence of several intracellular compartments (or organelles) with separate enzymes and molecules moving between them.

The field of bio-inspired computation has a long history marked by famous pioneers, such as Walter Pitts, Warren McCulloch, Alan Turing, Norbert Wiener, John von Neumann. In his unfinished book, *The Computer and the Brain* (first published in 1958), von Neumann presented his view on the analogies and difference between brains and computers and gave some directions for further research. Natural computing is now a very active research area that is concerned with the observation and study of computing-like processes taking place in nature and with the implementation of algorithms, programs and technical systems inspired by these natural processes. We can mention here a plethora of examples of systems and algorithms: cellular automata, evolutionary algorithms, neural networks, quantum computing, DNA computing, artificial immune systems, systolic arrays, artificial tissues, artificial life, synthetic genomes, particle swarm optimization, ant colony optimization, artificial bee colony algorithm, bacteria foraging, leaping frog algorithm, cuckoo search, bat algorithm, firefly algorithm, flower pollination algorithm, artificial plant optimization algorithm, bird flocking, fish schooling, protein memories, etc. These "many facets of natural computing" are investigated in a review by Kari and Rozenberg (2008) that takes the reader from the basic paradigms of cellular automata, artificial neuron model, and evolutionary computation, to the more recent concepts of swarm intelligence, artificial immune systems, and membrane computing.

A better understanding of the cell architecture and functioning has led to the development of new computing models. For example, DNA computing has emerged since 1994 with the seminal paper of Adleman as a fruitful and exciting new research field at the intersection of biology, computer science, mathematics, and engineering. Since then, various models of molecular computation have been proposed (filtering models, splicing models, constructive models, membrane models). Of these models, this book is focused on membrane computing which has been introduced in 1998 by the Romanian mathematician Gheorghe Păun. He proposed a parallel and distributed computing model starting from the study of cellular membranes. In 2003, the Institute for Scientific Information (ISI) considered membrane computing as an "emerging research area of computer science".

Membrane computing has attracted at the beginning mainly mathematicians and theoretical computer scientists. Lately, scientists from other disciplines have become interested in the use of this computing model, and we can mention here economists, applied computer scientists, engineers, and robotics specialists. Hundreds of researchers from more than 30 countries have been involved both in theoretical studies and in proposing and developing far reaching applications. There have been reported successful applications in a wide range of areas, such as linguistics, modelling of biological processes (systems and synthetic biology), computer graphics, optimization, economics, and multiple engineering areas. More than 2500 publications and 100 Ph.D. theses regarding membrane computing have been published by authors around the world since the introduction of the original model by Păun. In 2016, the International Membrane Computing Society has been founded with the goal of promoting research and cooperation within the vibrant membrane computing community.

Recently, a great interest has been shown in controlling mobile robots using membrane computing based approaches, a research area that has been sparked by the research done at the *Natural Computing and Robotics Laboratory* of the *Politehnica University of Bucharest*.

Robotics is a fast developing field with far reaching implications in every domain of our life. Much more, as the recent developments in the field show, robotics is a driving and integrative technology that will ultimately find a place in everyday life.

A reference document for the robotics professionals is the *Robotics 2020 Multi-annual Roadmap* where a detailed presentation of applications, abilities, and technologies for robotic systems is given. The application areas for robotic systems are very diverse: manufacturing, healthcare, agriculture, civil domain, commercial domain, logistics and transportation, consumer robots. Each of these broad areas can be further detailed on further sub-domains. For example, the consumer robots can be further classified into four sub-domains: domestic appliances, entertainment, education, and assisted living, each with their own objectives and constraints. For example, robots co-existing with children in domestic and public environments are expected to behave as safe, fun, and reliable companions and tutors, and be also capable of engaging in playful and meaningful interaction.

The robotic systems can possess (at different levels) various abilities, such as: adaptability, cognitive ability, configurability, decisional autonomy, dependability, interaction ability, manipulation ability, motion ability and perception ability.

Robotic technologies are grouped by purpose into four clusters, according to the same roadmap mentioned above:

1. Systems Development (systems engineering, system design, system integration): better systems and tools;
2. Human Robot Interaction (human machine interface, safety, human robot collaboration): better interaction;
3. Mechatronics (actuators, control, sensors, communication): making better machines; perception;
4. Navigation and Cognition (interpretation, sensing, motion planning, mapping, localization, natural interaction, cognitive architectures, action planning): better action and awareness.

Multi-robot systems are groups of homogeneous or/and heterogeneous robots that act in a shared environment in order to achieve tasks collectively when the given task is too complicated or impossible to achieve for a single robot, when the task is by its nature distributed, when we want the task to be fulfilled quicker, or when robustness is a key factor (the failure of a single robot would not make the task not achievable when there are multiple, redundant, robots in the group). The control in multi-robot systems is implemented in a centralized or decentralized manner. Multi-robot systems have found their way in a large number of applications, such as exploration and surveillance, search and rescue, warehouse management and operations planning (Kiva robots for the Amazon warehouses), robot toys and intelligent transportation systems. Swarm robotics is an area of multi-robot systems based on the principles of local interaction (possibly mediated by the environment) between simple robots that ultimately leads to the emergence of a complex macroscopic behavior. Swarms of flying robots can be used in disaster areas to create communication areas or in conflict areas to monitor adverse actions, while for example swarms of robotic fish with appropriate chemical sensors can be used for searching and monitoring contaminants and pollutants.

Networked robotics refers to the system of systems approach of robots, sensors, computers and users interconnected via network communication technologies. Robotic swarms are large multi-robot systems (typically homogeneous in what regards the morphology, size, behavior, cognition) based on inspiration from natural societies (ants, birds, bees) and on the principles of swarm intelligence: decentralized control and simple local control rules (interaction) between simple agents. They are mostly used as demonstrators of simple yet powerful interaction mechanisms that lead to interesting global behaviors that mimic the abilities of large colonies of simple insects in tasks like: search, foraging, grazing, harvesting, deployment, coverage, transportation, exploration, pursuit, predator-prey, etc. In the same time, examples of effective application of swarm robots have started to appear in the mainstream, such as the swarms of flying robots for search in disaster areas, inspection and maintenance with swarms of micro aerial robots, ocean monitoring, swarms of robotic fish.

Bio-inspired techniques have found numerous applications in robot design and control. Various innovative concepts and technologies have been designed and tested such as: robots with flapping wings, insect-like eyes, very sensitive robotic skins, artificial muscles, fish robots with chemical sensors, locust-like jumping mechanisms, soft actuators and sensors, conformable robots, climbing robots, helical swimming robots, robots with bio-inspired spine mechanisms, (high speed) running robots, robots with compliant pockets, flexible and transient electronics.

In the area of planning, optimization and control, there have been reported bio-inspired applications in the gait planning for biped robots, optimization of robot work cell layout, robust adaptive control of robot manipulators, robust control of robot arm-and-hand system, autonomous landing for flying robots, learning motion trajectories, fuzzy and neural controllers for mobile robots, multi robot area exploration, etc.

CONTENTS

We propose a systematic and detailed treatment of the use of membrane computing for robot control in a book which is easy to follow and clearly structured. We provide the fundamental concepts of membrane computing and mobile robotics and then we continue with methods for designing and testing membrane computing based controllers for single mobile robots as well as for swarms of mobile robots. We also approach the overlooked issue of robot swarms security using a membrane computing approach. Overall, our book provides a ready to use conceptual framework for how to integrate membrane computing in mobile robotics in addition to other modelling and simulation tools and computational approaches. There are many simulation examples, and part of the included experiments are validated on open-source, low cost mobile robots (*Kilobot*).

The dedicated webpage (http://membranecomputing.net/IGIBook/) is an additional source of information where source code, input files, user manuals, and demonstration videos can be found.

We now describe in detail the structure of this book.

Chapter 1 gives an overview of swarm robotics. First, we position this area in the context of the fast developing field of robotics and secondly, we present the fundamental concepts of decentralized control, local interaction, stigmergy, and emerging intelligence. Examples of typical robots used in swarm robotics experiments are given together with examples of robot simulators.

In Chapter 2 we introduce the fundamental concepts of membrane computing and define the basic P systems models that are considered relevant – at the time of this writing – to the robotics specialists. Among these are the numerical P systems, enzymatic numerical P systems, P colonies, XP colonies, and P swarms. Detailed examples will help the reader to understand how these systems work. We close this chapter with an extended analogy between membrane computing and swarm robotics that lies at the basis of our work reported in Chapter 4.

Chapter 3 is concerned with an overview of the software simulators for membrane computing. First we present some of the simulation approaches and software described in the literature, and then we continue with a more detailed presentation of the software simulators we have designed at the *Natural Computing and Robotics Laboratory* from the *Politehnica University of Bucharest*: *SNUPS* (for simulating standard and enzymatic numerical P systems), *Lulu* (an open-source simulator for P colonies and P swarms) and *PeP* (an open-source simulator for standard numerical P systems and their variants).

The core of this book is represented by Chapter 4. There we will describe in a very detailed manner how to use basic membrane computing models for controlling single and multiple robots, both simulated and real-world robots. Source code and state diagrams will intermediate a better understanding of the basic principles of membrane computing based control of robots.

In Chapter 5 we conclude this book by giving an overview of how to use membrane computing in robot control. We indicate which membrane computing models can be used to solve various problems in robotics, such as localization, obstacle avoidance, dispersion etc. We end with providing some research directions that could lead to the development of this field.

TARGET AUDIENCE

We hope that our book will be welcomed by a wide range of readers interested in understanding the fundamentals of membrane computing, and of controlling mobile robots using bioinspired approaches based on membrane computing models.

We will gladly welcome any comments, suggestions, and proposals for cooperation.

Andrei George Florea
Bucharest, Romania

Cătălin Buiu
Bucharest, Romania

Acknowledgment

The original research described in this book was supported by the grant of the Romanian Ministry of National Education CNCS-UEFISCDI, project number PN-II-ID-PCE-2012-4-0239, *Bioinspired techniques for robotic swarms security* (Contract #2/30.08.2013).

Chapter 1
An Overview of Swarm Robotics

ABSTRACT

Robotic swarms represent the application target of the studies presented in this book and therefore required the reader to be acquainted with the main concepts behind this branch of robotics. The introduction of swarm robotics principles is done only after presenting multi-robot systems, in comparison with single robot systems. Among the concepts that are defined in this chapter we mention: swarm robotic system, stigmergy and neighborhoods. After this theoretical introduction, the chapter continues with a presentation of robotic platforms that can be used to validate swarm algorithms. Among the robots listed are the Kilobot, the e-puck and the Khepera. As swarm robotics generally requires a large number of individuals, the costs of running experiments on real robots can become high. For this reason, robot simulation platforms are also discussed at the end of this chapter.

INTRODUCTION

Swarm robotics is a recent and important paradigm which is based on the principles of swarm intelligence applied to large groups of simple robots. Robotics itself is a fast developing area in which theoretical scientifical advances have been backed by impressive technological developments. The International Federation of Robotics (IFR press release, n.d.) is releasing every year robot statistics and forecasts. For 2015, it was reported the largest number of industrial and service robots sold. More than 41,000 units of professional service robots were sold with a sales value over 4.6 billion USD. As for the personal and domestic use service robots, there

DOI: 10.4018/978-1-5225-2280-5.ch001

were sold more than 5.4 million units with a sales value of over 2.2 billion USD. The projections for the period 2016-2019 indicate that more than 333,000 units of service robots for professional use and over 42 million units of service robots for personal and domestic use will be installed. The general industry has been the key sector leading the increasing demand for industrial robots which in 2015 increased by 15% with a global sales value of over 11.1 billion USD. At the end of 2015 there were 1.6 million operational industrial robots. Robot density is a key indicator when comparing the distribution of industrial robots in various countries. This indicator is defined as the number of multipurpose industrial robots per 10,000 persons employed in the manufacturing industry. The average robot density in 2015 was 69 (IFR press release, n.d.)

The *Robotics 2020 Multi-Annual Roadmap* issued by SPARC (Lafrenz, 2016) supports a Strategic Research Agenda in robotics. The roadmap provides an interesting and useful view of the robotics market with key market domains (manufacturing, healthcare, agriculture, consumer, civil, commercial, logistics and transport) and identifying key abilities for robots (Adaptability, Cognitive Ability, Configurability, Decisional Autonomy, Dependability, Interaction Ability, Manipulation Ability, Motion Ability and Perception Ability).

Swarm robotics is based on collective intelligence that arises from the interaction of many, simple robots. While most of the experiments in swarm robotics have been performed in simulation and to demonstrate key macroscopic behaviors, the real-world applications of swarm robotic systems are gaining momentum. The goal of this chapter is to provide an introduction to the principles of swarm robotics, key definitions, some examples of robots and software platforms that are commonly used in swarm robotics experiments.

SWARM ROBOTICS

Multi-Robot Systems: An Overview

Multi-robot systems (or collective robotic systems) are systems composed of multiple (autonomous) mobile robots and they offer important advantages as compared to single robot systems (Parker, 2008): (1) a task may be too complex for a single robot to achieve; (2) when the task by itself is distributed this requires the use of multiple robots to achieve it; (3) a single very complex robot might be more expensive to build than many simpler robots; (4) the inherent parallelism of a multi-robot system can help to solve the task quicker; (5) robustness is greatly improved when using multiple robots. There are different ways in which a multi-robot control architecture can be organized: centralized, hierarchical, decentralized, and hybrid.

Networked robots are a special case of multiple robot systems in that the coordination and cooperation between individual agents (robots) is achieved through networked communication. According to the IEEE RAS Technical Committee on Networked Robots (http://www-users.cs.umn.edu/~isler/tc/), networked robots are robotic devices connected to a communications network and can be divided into: (1) tele-operated robots (commands and feedback are exchanged through the network with human operators) and (2) autonomous networked robots (robots and sensors exchange data through the network).

Swarm robotics is an expanding research area in collective robotics that studies how emergent behaviors arise from direct local interactions between a large number of simple robots and through their indirect interaction via a shared environment. There is no central coordination of the robots, yet the swarm, as a whole, displays capabilities that are beyond those of individual robots. The main motivations for swarm robotics research are scalability (adding any number of additional robots would not change the group behavior as a whole), flexibility (the swarm is capable to operate with success in different environments while performing different tasks), and robustness (failure of single robots would not affect the behavior and performance of the swarms as a whole) (Şahin, 2005).

A large number of macroscopic swarm behaviors have been studied including: aggregation, flocking, foraging, natural herding, schooling, sorting, clumping, aggregation, containment, orbiting, surrounding, perimeter search, evasion, tactical overwatch, path formation, condensation, dispersion, search, grazing, harvesting, deployment, coverage, localization, mapping, exploration, pursuit, predator-prey, target tracking. For detailed reviews of typical swarm behaviors, see (Bayındır, 2015) and (Parker, 2008). Some of these behaviors are considered canonical, such as homing, dispersion, clustering, and monitoring (Duarte et al., 2016)

The control in a swarm robotic system is typically decentralized. Given a desired global behavior of the swarm, how to generate the behavioral rules that determine the interactions between robots is a difficult problem, for which there is no general solution (Dorigo, Trianni, Şahin, Groß, & Labella, 2004). Studying and devising control laws for swarms often require an interdisciplinary approach. In (Bouffanais, 2015) there are reviewed five possible approaches to collective behaviors, swarming, and collective dynamics, namely: (1) the biologically inspired approach, (2) the physical approach, (3) the network-theoretic approach, (4) the information-theoretic approach and (5) the computational approach.

Most of the studies in swarm robotics have been conducted on simulated robots, and only a few have been conducted on real-world setups. Additionally, most of the experiments with real robots have been performed in controlled laboratory settings. However, recently, the interest in swarms of aerial or marine swarm robotic systems has increased. For example, in (Duarte et al., 2016) a study on the efficiency of evolutionary control of a swarm of aquatic robotics is performed. Human-swarm interaction is also an active area of research, see e.g. (Kolling, Walker, Chakraborty, Sycara, & Lewis, 2016).

An overlooked issue in designing swarm robotics systems is security. An early review of the main security hazards for robotic swarms is given in (Higgins, Tomlinson, & Martin, 2009). A further discussion can be found in (Buiu & Gansari, 2014), and in Chapter 4 we present a bioinspired approach to the security of robotic swarms using membrane computing.

An excellent review of swarm robotics from the swarm engineering perspective is given in (Brambilla, Ferrante, Birattari, & Dorigo, 2013).

Swarm Robotics Principles

A swarm robotic system has taken its inspiration from societies of biological agents. In nature there are systems of biological agents which can collectively solve tasks that are far beyond the capabilities of a single biological agent. Standard examples are the social insects (ant, termite, bee, wasp colonies), fish school or bird flocks. For example, an ant colony solves complex tasks, such as transportation of large objects (prey) or building complex structures, like nests. For example, a colony of leaf-cutter ants can have up to millions of individuals and is considered to have the most elaborate communication system, the most elaborate caste, and the most sophisticated nest and can be considered a superorganism by itself (Hölldobler & Wilson, 2008).

Definition 1: Swarm Robotic System

A swarm robotic systems (SRS) is a super-additive multi-robot system with a large number of simple robots with no central controller. The swarm's global behavior emerges because of direct and indirect interactions between robots.

Typically, there are hundreds or thousands of simple robots in a swarm. Usually, A SRS is homogeneous, i.e. all the robot are morphologically and functionally identical. However, there are cases when the robots in a SRS have different physical and software abilities, i.e. the SRS becomes heterogeneous. There are direct local interactions between robots and indirect interactions which are mediated by the shared environment. Global behaviors emerge as a result of these interactions. So, the SRS can be considered super-additive, in the sense that the whole is much bigger than the sum of the parts.

Definition 2: Stigmergy

Stigmergy is a universal coordination mechanism (Heylighen, 2016a, 2016b; Susi, 2016) that can be applied to the analysis of insect societies, human societies, and web communities. In the context of SRSs, stigmergy refers to the indirect coordination of robots' actions via the traces they leave in the environment and which stimulate the further actions of the agents (robots) that "find" these traces.

This mechanism can be implemented in robotic swarms in a number of ways, see for example the concept of extended stigmergy that was introduced in (Werfel & Nagpal, 2006).

Definition 3: Subswarms and Neighborhoods

Continuing interactions between robots in a swarm can lead to formation of dynamic, adaptive, subswarms and neighborhoods. Subswarms (subgroups of robots) are formed inside the swarm in order to accomplish specific tasks. Neighborhoods (smaller groups of robots) can be formed inside a subswarm in order for the member robots to help each other. Both subswarms and neighborhoods can be created or modified in a static or dynamic way.

For example, when there are multiple goals to achieve, or the task itself has a distributed nature, subswarms can be formed and different goals will be assigned to different subswarms. When considering a given goal from the overall set of goals, neighborhoods can be formed in order for the member robots to accomplish their common goal and to improve the solving capabilities of a subswarm. An example of a swarm structure where subswarms and neighborhoods have been formed is given in Figure 1a. A tree-like representation of this architecture is given in Figure 1b. Note the similarity with the membrane structure of a standard P system as presented in Chapter 2 (Figure 1a. and 1b).

ROBOTS FOR SWARM ROBOTICS EXPERIMENTS

The robots in a typical robotic swarm must be simple agents with limited sensing and acting capabilities. The *Kilobot* robot (Figure 2) is a low-cost robot designed to allow development and testing of control algorithms for swarms of hundreds and thousands of robots (Rubenstein, Ahler, Hoff, Cabrera, & Nagpal, 2014). We used this robot, both in simulation and real-world settings, for our experiments in membrane computing based control of robotic swarms which are described in Chapter 4. This robot has three rigid legs, two vibration motors, an ambient light sensor, a three-color LED, and an infrared transmitter/receiver used for a robot-to-robot distance sensing

Figure 1. A homogeneous SRS with 29 robots. a. The dynamic hierarchical structure of a swarm: the outer border (1) encloses the whole swarm in which two subswarms have formed, delimited by borders 2 and 6. In each subswarm a number of neighborhoods can be noticed (3, 4, 5 in one subswarm and 7, 8, 9, 10 in the other subswarm) b. A tree-like representation of the swarm

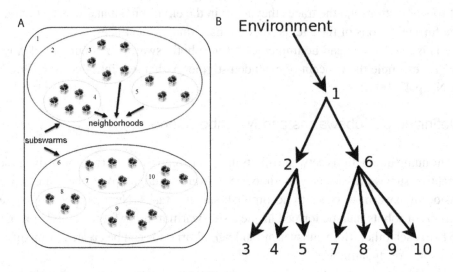

mechanism. It has no distance sensors and so it cannot detect (and avoid) obstacles. Typical *Kilobot* swarm robotics experiments involved self-organization (Rubenstein et al., 2014), collective decision (Valentini, Ferrante, Hamann, & Dorigo, 2015), and supervisory control (Lopes, Leal, Dodd, & Groß, 2014).

Another widely used platform for swarm robotics experiments is the e-*puck* robot, a 7 cm differential wheeled robot (Figure 3), an open-hardware and open-software robot. Open refers here to open source, a term that has a similar meaning for both software and hardware. Historically, the first to appear was open source software as a generic license term for software that is freely available not only as a compiled product but also includes access to its source code (Laurent, 2004), The term open source also refers to the freedom to modify, redistribute and integrate the product into other applications. Open source hardware builds on the foundations of open source software in that it refers to hardware "whose source files are publicly available for anyone to use, remanufacture, redesign and resell" (Gibb, 2014). It features infrared proximity sensors, an onboard camera, an accelerometer, a ground sensor, three sound sensors, a loudspeaker, and various LEDs for communications. Four case studies which use the e-*puck* and *Kilobot* robots and experiments with up to 600 physical robots based on the principles of supervisory control theory are reported in (Lopes, Trenkwalder, Leal, Dodd, & Groß, 2016).

Figure 2. a. A group of Kilobots,b. A simulated Kilobot,c. The components of a Kilobot (three metal legs (A), RGB LED (B), vibrating motors (C), light sensor (D), charging hook (E), battery (F), infrared emitter-receiver (G)

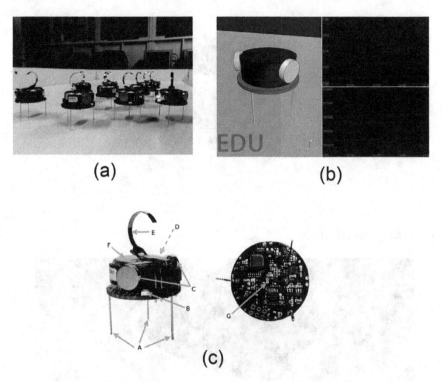

(a)

(b)

(c)

The *Khepera* robot has been developed since 1991 (Mondada, Franzi, & Guignard, 1999) and is widely used as a tool for testing control algorithms on single robot systems and on robotic swarms. *Khepera IV*, the latest in its series, improves over its predecessor, *Khepera III* (Figure 4), in what regards the odometry, and features a powerful *Linux* core with Wi-Fi, Bluetooth, accelerometer, gyroscope, and color camera (Kteam, 2016). Multi-robot path planning is designed and optimized on a group of *Khepera* robots in (Das, Behera, & Panigrahi, 2016) and an evolutionary optimization based method for solving a multi-robot stick-carrying problem with simulated and real *Khepera* robots is discussed in (Sadhu, Rakshit, & Konar, 2016). Figure 5 shows a group of 10 e-puck robots and one *Khepera III*.

Figure 3. e-puck robot

Figure 4. Khepera III robot

ROBOT SIMULATION PLATFORMS

There are numerous software packages available for simulating robots. We will briefly review only a few of them. *Webots* is a development environment for modelling, programming and simulating mobile robots (Webots, n.d.). The developed programs can be transferred to commercially available real robots (Michel, 2004), such as *e-puck*, *Koala*, *Khepera*, or *Nao* (Michel, 2004). The current version is 8.5.0.

V-REP features an integrated development environment (Coppelia Robotics V-REP: Create. Compose. Simulate. Any Robot., n.d.) and is based on a distributed

Figure 5. A heterogeneous swarm of 11 robots (10 e-puck and 1 Khepera III)

control architecture (Rohmer, Singh, & Freese, 2013). *V-REP* can be easily installed on *Linux*, *OSX*, and *Windows* simply by uncompressing a zip archive. The controllers can be developed in C/C++, Python, Java, Lua, Matlab, Octave or Urbi and interfaced with the simulator using dedicated libraries that are provided for each of the external languages. These external libraries interact with a specified robot from the currently simulated scene where each robot runs a Lua script that is also programmed by the user to respond to various commands that are sent by the external library. The latest version is 3.3.2. A view of the *V-REP* main window is presented in Figure 6 where one can notice the graphical user interface of the simulator and a highlight of the most important components. The most important component is the 3D simulation frame (1) that allows user interaction (camera and robot selection, translation, rotation) using the mouse. The tree view on the left (2) allows the user to select subcomponents of objects such as robot sensors, and alter their settings. The simulation is controlled using the toolbar (3) on the top part of the interface and once started, the evolution of one or more robots can be viewed and recorded by means of graphs that use a user specified variable (X and Y positions in (4) and battery level in (5)). The interface also features a console (6) that allows for the printing of output messages from user scripts written in Lua that are executed by the simulated robots.

Figure 6. V-REP IDE main view that is composed of: (1) 3D simulation frame, (2) simulation scene object tree, (3) simulation control toolbar, (4) X and Y position / time graph, (5) battery level / time graph, (6) Lua console

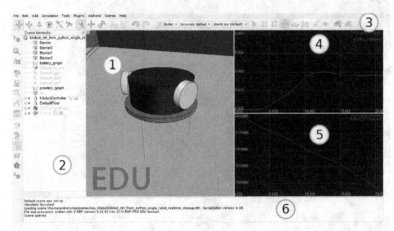

A *Khepera IV* library for *V-REP* was designed using *Autodesk Inventor* for the visual interface and Lua as programming language. This library is proposed and described in (Peralta, Fabregas, Farias, Vargas, & Dormido, 2016). *Kilombo* is a powerful robot simulator which greatly facilitates the simulation of large swarm robotic systems (Jansson et al., 2015). Further details and an application are given in Chapter 3 of this book. *ARGoS* is a multi-physics robot simulator (The ARGoS Website, n.d.) which can be used to simulate large robotic swarms (Pinciroli et al., 2012). It is also an option to simulate *Kilobot* robots. ROS code can also be run on *ARGoS*. ROS is the acronym for *Robot Operating System*, a meta operating system which facilitates the development of robot software through the use of a collection of software frameworks and is becoming a de facto standard in robotics. In (Mendonca et al., 2013) there is described *Kelpie*, a ROS-base multi-robot simulator for water, surface and aerial vehicles. *Gazebo* is a 3D simulator often used with ROS, see for example (Farinelli, Boscolo, Zanotto, & Pagello, 2016) and (Rosa, Cognetti, Nicastro, Alvarez, & Oriolo, 2015). *Gazebo* extends Stage, a C++ library that simulates multiple robot systems (Vaughan, 2008), while *Stage* itself is interfaced to *Player*, a widely used open source simulation platform in robotics (Collett, MacDonald, & Gerkey, 2005). *RobotNetSim* is a simulation framework for networked multi-robot systems (Kudelski, Gambardella, & Di Caro, 2013). For other simulation platforms and robots used in swarm robotics experiments, the interested reader may consult (Tan & Zheng, 2013).

CONCLUSION

In this chapter, we have provided a brief introduction to swarm robotics. We discussed about local interactions, stigmergy, and emergent behaviors. Possible spatial and logical formations inside of a swarm robotic systems were discussed. We have given a list of robots that are typically used in swarm robotics experiments (*Kilobot, e-puck*, and *Khepera*) and a list of robot simulators that are commonly used to simulate robotic swarms.

REFERENCES

Bayındır, L. (2015). A review of swarm robotics tasks. *Neurocomputing, 172*, 292–321. doi:10.1016/j.neucom.2015.05.116

Bouffanais, R. (2015). *Design and Control of Swarm Dynamics*. Singapore: Springer Singapore; doi:10.1007/978-981-287-751-2

Brambilla, M., Ferrante, E., Birattari, M., & Dorigo, M. (2013). Swarm robotics: A review from the swarm engineering perspective. *Swarm Intelligence, 7*(1), 1–41. doi:10.1007/s11721-012-0075-2

Buiu, C., & Gansari, M. (2014). A new model for interactions between robots in a swarm. In *Electronics, Computers and Artificial Intelligence (ECAI), 2014 6th International Conference on* (pp. 5–10). doi:10.1109/ECAI.2014.7090202

Collett, T., MacDonald, B., & Gerkey, B. (2005). Player 2.0: Toward a practical robot programming framework. In *Proceedings of the Australasian Conf. on Robotics and Automation (ACRA.* Das, P. K., Behera, H. S., & Panigrahi, B. K. (2016). Intelligent-based multi-robot path planning inspired by improved classical Q-learning and improved particle swarm optimization with perturbed velocity. *Engineering Science and Technology, an International Journal, 19*(1), 651–669. http://doi.org/doi:<ALIGNMENT.qj></ALIGNMENT>10.1016/j.jestch.2015.09.009

Dorigo, M., Trianni, V., Şahin, E., Groß, R., & Labella, T. (2004). *Evolving self-organizing behaviors for a swarm-bot*. Autonomous.

Duarte, M., Costa, V., Gomes, J., Rodrigues, T., Silva, F., Oliveira, S. M., & Christensen, A. L. (2016). Evolution of collective behaviors for a real swarm of aquatic surface robots. *PLoS ONE, 11*(3), e0151834. doi:10.1371/journal.pone.0151834 PMID:26999614

Farinelli, A., Boscolo, N., Zanotto, E., & Pagello, E. (2016). Advanced approaches for multi-robot coordination in logistic scenarios. *Robotics and Autonomous Systems*. doi:10.1016/j.robot.2016.08.010

Gibb, A. (2014). *Building Open Source Hardware: DIY Manufacturing for Hackers and Makers*. Addison-Wesley Professional.

Heylighen, F. (2016a). Stigmergy as a universal coordination mechanism I: Definition and components. *Cognitive Systems Research*, *38*, 4–13. doi:10.1016/j.cogsys.2015.12.002

Heylighen, F. (2016b). Stigmergy as a universal coordination mechanism II: Varieties and evolution. *Cognitive Systems Research*, *38*, 50–59. doi:10.1016/j.cogsys.2015.12.007

Higgins, F., Tomlinson, A., & Martin, K. M. (2009). Threats to the Swarm : Security Considerations for Swarm Robotics. *International Journal on Advances in Security*, *2*(2&3), 288–297. http://doi.org/10.1.1.157.1968

Hölldobler, B., & Wilson, E. O. (2008). *The Superorganism: The Beauty, Elegance, and Strangeness of Insect Societies*. W. W. Norton & Company.

IFR press release. (n.d.). Retrieved from http://www.ifr.org/news/ifr-press-release/service-robotics-835/

Jansson, F., Hartley, M., Hinsch, M., Slavkov, I., Carranza, N., & Olsson, T. S. G. ... Grieneisen, V. A. (2015). Kilombo: a Kilobot simulator to enable effective research in swarm robotics. *arXiv Preprint arXiv:1511.04285*.

Kolling, A., Walker, P., Chakraborty, N., Sycara, K., & Lewis, M. (2016). Human Interaction With Robot Swarms: A Survey. *IEEE Transactions on Human-Machine Systems*, *46*(1), 9–26. doi:10.1109/THMS.2015.2480801

Kteam (2016, November 29). BEYOND MINIATURE TECHNOLOGY. Retrieved from http://www.k-team.com/

Kudelski, M., Gambardella, L. M., & Di Caro, G. A. (2013). RoboNetSim: An integrated framework for multi-robot and network simulation. *Robotics and Autonomous Systems*, *61*(5), 483–496. doi:10.1016/j.robot.2013.01.003

Lafrenz, R. (6 June 2016). Multi-Annual Roadmap (MAR) for Horizon 2020 Call ICT-2017 (ICT-25, 27 & 28) published. Retrieved February 15, 2017, from https://www.eu-robotics.net/sparc/newsroom/press/multi-annual-roadmap-mar-for-horizon-2020-call-ict-2017-ict-25-ict-27-ict-28-published.html?changelang=2

Laurent, A. M. S. (2004). *Understanding open source and free software licensing: guide to navigating licensing issues in existing & new software*. O'Reilly Media, Inc.

Lopes, Y. K., Leal, A. B., Dodd, T. J., & Groß, R. (2014). Application of Supervisory Control Theory to Swarms of e-puck and Kilobot Robots. In *Swarm Intelligence* (pp. 62–73). Springer International Publishing; doi:10.1007/978-3-319-09952-1_6

Lopes, Y. K., Trenkwalder, S. M., Leal, A. B., Dodd, T. J., & Groß, R. (2016). Supervisory control theory applied to swarm robotics. *Swarm Intelligence, 10*(1), 65–97. doi:10.1007/s11721-016-0119-0

Mendonca, R., Santana, P., Marques, F., Lourenco, A., Silva, J., & Barata, J. (2013). Kelpie: A ROS-Based Multi-robot Simulator for Water Surface and Aerial Vehicles. In *2013 IEEE International Conference on Systems, Man, and Cybernetics* (pp. 3645–3650). IEEE. http://doi.org/ doi:10.1109/SMC.2013.621

Michel, O. (2004). Webots: Professional Mobile Robot Simulation. *International Journal of Advanced Robotic Systems, 1*(1), 39–42. doi:10.5772/5618

Mondada, F., Franzi, E., & Guignard, A. (1999). The Development of Khepera. *Experiments with the Mini-Robot Khepera, Proceedings of the First International Khepera Workshop*, 7–14.

News. (n.d.). Retrieved from http://www.coppeliarobotics.com/

Parker, L. E. (2008). Multiple Mobile Robot Systems. In *Springer Handbook of Robotics* (pp. 921–941). Berlin, Heidelberg: Springer Berlin Heidelberg; doi:10.1007/978-3-540-30301-5_41

Peralta, E., Fabregas, E., Farias, G., Vargas, H., & Dormido, S. (2016). Development of a Khepera IV Library for the V-REP Simulator. *IFAC-PapersOnLine, 49*(6), 81–86. doi:10.1016/j.ifacol.2016.07.157

Pinciroli, C., Trianni, V., OGrady, R., Pini, G., Brutschy, A., Brambilla, M., & Dorigo, M. et al. (2012). ARGoS: A modular, parallel, multi-engine simulator for multi-robot systems. *Swarm Intelligence, 6*(4), 271–295. doi:10.1007/s11721-012-0072-5

Rohmer, E., Singh, S. P. N., & Freese, M. (2013). V-REP: A versatile and scalable robot simulation framework. In *2013 IEEE/RSJ International Conference on Intelligent Robots and Systems (IROS)* (pp. 1321–1326). doi:10.1109/IROS.2013.6696520

Rosa, L., Cognetti, M., Nicastro, A., Alvarez, P., & Oriolo, G. (2015). Multi-task Cooperative Control in a Heterogeneous Ground-Air Robot Team. *IFAC-PapersOnLine, 48*(5), 53–58. doi:10.1016/j.ifacol.2015.06.463

Rubenstein, M., Ahler, C., Hoff, N., Cabrera, A., & Nagpal, R. (2014). Kilobot: A low cost robot with scalable operations designed for collective behaviors. *Robotics and Autonomous Systems*, *62*(7), 966–975. doi:10.1016/j.robot.2013.08.006

Sadhu, A. K., Rakshit, P., & Konar, A. (2016). A modified Imperialist Competitive Algorithm for multi-robot stick-carrying application. *Robotics and Autonomous Systems*, *76*, 15–35. doi:10.1016/j.robot.2015.11.010

Şahin, E. (2005). Swarm Robotics: From Sources of Inspiration to Domains of Application. In *Proceedings of the 2004 International Conference on Swarm Robotics* (pp. 10–20). Springer-Verlag. http://doi.org/ doi:<ALIGNMENT.qj></ALIGNMENT>10.1007/978-3-540-30552-1_2

Susi, T. (2016). Social cognition, artefacts, and stigmergy revisited: Concepts of coordination. *Cognitive Systems Research*, *38*, 41–49. doi:10.1016/j.cogsys.2015.12.006

Tan, Y., & Zheng, Z. (2013). Research Advance in Swarm Robotics. *Defence Technology*, *9*(1), 18–39. doi:10.1016/j.dt.2013.03.001

The ARGoS Website. (n.d.). Retrieved from http://www.argos-sim.info/

Valentini, G., Ferrante, E., Hamann, H., & Dorigo, M. (2015). Collective Decision with 100 Kilobots : Speed vs Accuracy in Binary Discrimination Problems. *IRIDIA – Technical Report Series*, (July).

Vaughan, R. (2008). Massively multi-robot simulation in stage. *Swarm Intelligence*, *2*(2–4), 189–208. doi:10.1007/s11721-008-0014-4

Webots. (n.d.). Retrieved from https://www.cyberbotics.com/overview

Werfel, J., & Nagpal, R. (2006). Extended Stigmergy in Collective Construction. *IEEE Intelligent Systems*, *21*(2), 20–28. doi:10.1109/MIS.2006.25

Chapter 2
Membrane Computing:
Theory and Applications

ABSTRACT

The theoretical computing models that are used throughout this book are described in this chapter. These models are based on the initial P system model and include: Numerical P systems, Enzymatic Numerical P systems, P colonies and P swarms. Detailed examples and execution diagrams help the reader allow the reader to understand the functioning principle of each model and also its potential in various applications. The similarity between P systems (and their variants) and robot control models is also addressed. This analysis is presented to the reader in a side-by-side manner using a table where each row represents an analysis topic. Among others we mention: (1) Architectural structure, (2) Modularity and hierarchy, (3) Input-output relationships, (4) Parallelism.

INTRODUCTION

P systems were introduced in a technical report from 1998 by the Romanian mathematician Gheorghe Păun and are at the foundation of membrane computing. The purpose of this chapter is to give an overview of the basic principles and models in membrane computing. Starting from analogies with biomembranes, the structure of a standard P system is presented. Then, several variants of the original model are discussed. The main paradigms that are relevant from the point of view of robotics applications, including standard numerical P systems, enzymatic numerical P systems, P colonies, and P swarms, are presented with examples.

DOI: 10.4018/978-1-5225-2280-5.ch002

Membrane computing is part of the larger field of molecular computing. DNA computing is perhaps the most known paradigm of molecular computing (Amos, Păun, Rozenberg, & Salomaa, 2002) and was initiated by the early paper of Adleman in which he described the use of DNA as a computational system (Adleman, 1994) and also implemented a relevant computation in a biological laboratory.

The models of molecular computation are categorized in (Amos, 2010) in four categories: filtering, splicing, constructive, and membrane-based

Filtering models are based on computations performed on finite multisets of strings. The concept of a multiset will be detailed in the discussion on P systems. For more details about filtering models, the interested reader is referred to (Amos, 2010). Splicing systems have been introduced in the seminal paper of Tom Head from 1987 (Head, 1987). The splice operation, given two strings S and T, is identical to the crossover operator employed by genetic algorithms (Holland, 1992). A restricted form of the splicing operation is used in the Parallel Associative Memory Model introduced in (Reif & H., 1995). Constructive models are based on the principles of self-assembly, and an example is the Tile Assembly Model introduced in (Rothemund & Winfree, 2000).

The most well-known class of membrane models is that of P systems, introduced by the Romanian mathematician Gheorghe Păun (Păun, 2000) and this will be described in the following part of this chapter. Next we will give a detailed analogy between P systems and swarm robotic systems that lays the foundations for our original contributions presented in Chapter 4.

P SYSTEMS

The Basic Model

In general terms, a system can be defined by a number of interdependent parts forming a unified whole. A system has well defined boundaries which separate it from the environment and other systems. Biological membranes make the cell a system as they separate the cell (and the cellular contents) from its external world. More, this separation is permeable in order to support the flow of nutrients to the cell and the flow of cellular waste outside the cell, and it is flexible in order to enable the cell to change form and to grow. It is demonstrated that all these membrane functions are due to the special physical properties of the molecules that build up the membranes (Phillips, Kondev, & Theriot, 2012). More, inside the cell there are organelles which are also bound by complex membranes (see for example, the nuclear envelope, the rough endoplasmic reticulum, and the mitochondrion).

Figure 1. a. A membrane structure for a standard P system b. Tree-like representation of the compartments

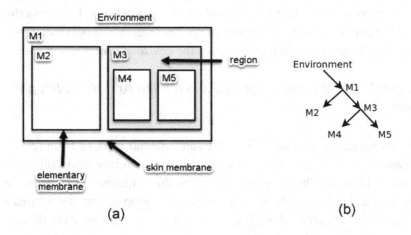

(a) (b)

P systems (P from Păun) were introduced (Păun, 2000) as membrane models of molecular computation inspired by the structure and roles of the biological membranes. Since the introduction of the basic distributed and parallel computing model (P system), there have been proposed and analyzed a large number of extensions which are, part of them, described in (Păun, Rozenberg, & Salomaa, 2010).

A membrane structure for a standard P system is an arrangement of compartments which can also be represented in a tree-like form as shown in Figure 1, where there are five membranes: M1 is the skin membrane (the outermost membrane), and the other four membranes (M2 and M3, on one hand, and M4 and M5, on the other hand, are siblings). The space inside a membrane is considered a region. M2 (M4 and M5 as well) is an elementary membrane as it has no descendant membranes.

Inside these compartments, there are multisets of objects and evolution rules which define how these multisets evolve. A multiset is a set which may contain more than one copy of an element, for example $a^2b^3c^5$ is a multiset in which there are 2 copies of a, 3 copies of b, and 5 copies of c. An evolution rule takes the general form $before \rightarrow after$, showing the transition of a multiset of objects (in the *before* state) to a new multiset of objects (the *after* state). For example, $a \rightarrow ab$ is a rule saying that the symbol a will transform itself in the pair ab (if there is a symbol a in the current compartment and the program can be applied). Special symbols may be used, e.g. the δ symbol in the evolution rule $a \rightarrow b\delta$, which translates as "if there is an a in the current compartment, transform it into b, and then dissolve (this is the meaning of δ) the current membrane". Dissolving a membrane means to eliminate the boundaries of the compartment whose membrane is being dissolved and to place the contents in the parent compartment. The rules are applied in a nondeterministic, maximally parallel way: assign objects to rules non-deterministically in parallel, in a maximal way, that is, until no further assignment of objects to rules are possible. When there are no rules that can be applied, the computation halts, and the final result of the computation is the number of special objects sent out through the most exterior membrane (the skin membrane) to the environment.

Example 1: Computing n^2 for Any $n \geq 0$ With Any n≥0 With a P System

The following example is taken from the fundamental work of Păun (2000). The membrane structure, the initial multisets of the objects and the sets of rules are given in Figure 2. There are three regions defined by three membranes (1 being the skin membrane), and there are objects only inside the region defined by the membrane 3 (objects a and f). When there is a ">" sign in an evolution rule, this shows a priority in executing that rule. For the rule $(ff \rightarrow f) > (f \rightarrow \delta)$, this means that the first part of the rule $ff \rightarrow f$ has priority over the second part of the rule $f \rightarrow \delta$, in other words, halve the number of f symbols in that region until there remains a single f, which then transforms itself in a δ, the special symbol that indicated that the current membrane will be dissolved. The rules in the third region (defined by membrane 3) are iterated in parallel for n times (n is the input to this P system, the number which we want to square). In Figure 3 there is shown in a graphical way the evolution of the P system until the execution halts.

A P system can be formally described using the following definition.

Definition 1: Standard P System (A Cell-Like P System With Multiset Rewriting Rules).

A cell-like P system with multiset rewriting rules of degree $m \geq 1$ is a construct:

$$\Pi = \left(O, H, \mu, w_1, \ldots, w_m, R_1, \ldots, R_m, i_0\right) \tag{1}$$

where O is an alphabet of objects, H is the alphabet of membrane labels, μ is a membrane structure of degree m, $w_1, \ldots, w_m \in O^*$ are the multisets of objects from the m regions of μ, R_i are the rewriting rules associated with each of the m regions ($1 \leq i \leq m$), and $i_0 \in H \cup \{e\}$ specifies the input/output region of Π.

A P system can be considered a computing device: the number of objects in a certain membrane can be considered the input of the system, and the number of objects in a specified membrane at the end of the computation is considered the output

Figure 2. An example P system

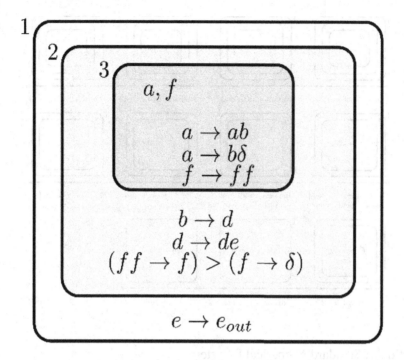

of the system. While P systems have been extensively analyzed in terms of their mathematical properties, most P systems variants are computationally universal and computationally efficient and there have been numerous applications (summarized in a sub-chapter below). In what regards the physical realization of a P system, there are only several prototypes as the one in (Nguyen, Kearney, & Gioiosa, 2010).

Numerical P Systems

The numerical P systems are considered, together with spiking neural P systems, as an "exotic" type of P systems as their degree of similarity to the functioning of a real cell is lower than in the case of standard P systems. However, there has been a large interest both in the computational properties of both types of P systems mentioned above, and in their practical applications. For example, the interest in numerical P systems has been largely stimulated by their first use in the development of membrane controllers for mobile robots (Buiu, Vasile, & Arsene, 2012).

Compared to the standard P system defined above, a numerical P system is using *numerical variables* and *programs* for evolving variables.

Figure 3. The evolution of the transitional P system from example 1

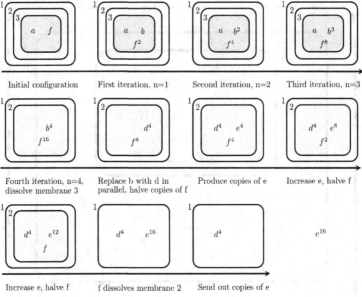

Definition 2: Standard Numerical P System.

A numerical P system is a construct of the following form (Păun & Păun, 2006):

$$\Pi =$$
$$\left(m, H, \mu, \left(Var_1, \ Pr_1, \ Var_1(0), Var_2, \ Pr_2, \ Var_2(0), ..., Var_m, \ Pr_m, \ Var_m(0)\right)\right) \qquad (2)$$

where m is the number of membranes (at least 1), H is an alphabet with m symbols (labels for membranes), μ is a membrane structure (showing the interconnection relationships between membranes), Var_i is the set of variables associated to compartment i ($Var_i = \left\{x_{1,i}, ..., x_{k_i,i}\right\}$), Pr_i is the set of programs from the same compartment i, and $Var_i(0)$ is the set of initial values for the variables in compartment i.

Each program l from compartment i has the following form:

$$Pr_{l,i} = \left(F_{l,i}\left(x_{1,i}, ..., x_{k_i,i}\right), c_{l,1} \middle| v_1 + c_{l,2} \middle| v_2 + ... + c_{l,n_i} \middle| v_{n_i}\right) \qquad (3)$$

The sum of the coefficients of program l in compartment i is denoted by $C_{l,i} = \sum_{s=1}^{n_i} c_{l,s}$ and each variable $v_{l,s}$ will receive a contribution $q * c_{l,s}$, $1 \leq s \leq n_i$, where $q = F_{l,i}\left(x_{1,i}(t), \ldots, x_{k_i,i}(t)\right) / C_{l,i}$.

At each time step, a program from each compartment is chosen non-deterministically for execution, the value of the production function is computed and according to the repartition protocol the variables' values at the next time step are computed.

Example 2: A Deterministic Numerical P System

The following example of a deterministic standard numerical P system (one program per each compartment) is taken from (Păun & Păun, 2006):

$$\Pi = (4, H, \mu, (Var_1, \ Pr_1, \ Var_1(0), Var_2, \ Pr_2, \ Var_2(0),$$

$$Var_3, \ Pr_3, \ Var_3(0), Var_4, \ Pr_4, \ Var_4(0)) \tag{4}$$

where:

$$\mu = [_1[_2[_3]_3[_4]_4]_2]_1$$

$Var_1 = \{x_{1,1}\},$
$Pr_1 = (2x_{1,1}^2, 1 \mid x_{1,1} + 1 \mid x_{1,2}),$
$Var_1(0) = (1),$

$Var_2 = \{x_{1,2}, x_{2,2}, x_{3,2}\},$
$Pr_2 = (x_{1,2}^3 - x_{1,2} - 3x_{2,2} - 9, 1 \mid x_{2,2} + 1 \mid x_{3,2} + 1 \mid x_{2,3}),$
$Var_2(0) = (3, 1, 0) ,$

$Var_3 = \{x_{1,3}, x_{2,3}\},$
$Pr_3 = (2x_{1,3} - 4x_{2,3} + 4, 2 \mid x_{1,3} + 1 \mid x_{2,3} + 1 \mid x_{1,2}),$
$Var_3(0) = (2, 1) ,$

$$Var_4 = \{x_{1,4}, x_{2,4}, x_{3,4}\},$$
$$Pr_4 = (x_{1,4}x_{2,4}x_{3,4}, 1 \mid x_{1,4} + 1 \mid x_{2,4} + 1 \mid x_{3,4} + 1 \mid x_{3,2}),$$
$$Var_4(0) = (2,2,2),$$

A graphical representation of this deterministic numerical P system can be seen in Figure 4. A detailed step-by-step evolution of this P system is described in (Păun & Păun, 2006).

In (Pavel, Arsene, & Buiu, 2010) there was introduced an extension of numerical P systems based on an analogy with biological enzymes. The concept was further extended and exemplified in various robot control applications in (Pavel & Buiu, 2011; Pavel, Vasile, & Dumitrache, 2012).

The definition of an enzymatic P system differs from one of a standard numerical P system only in the structure of programs. In this case, there can be multiple production functions in the membranes and some special variables called enzymes are involved only in the selection of the valid production functions. More, as their biological counterparts (real enzymes), they are not consumed in the production functions (in the biochemical reactions they catalyze in a cell, respectively). All the production functions that fulfill the execution condition will be executed in parallel.

Figure 4. An example of a numerical P system with four membranes and one production function per compartment

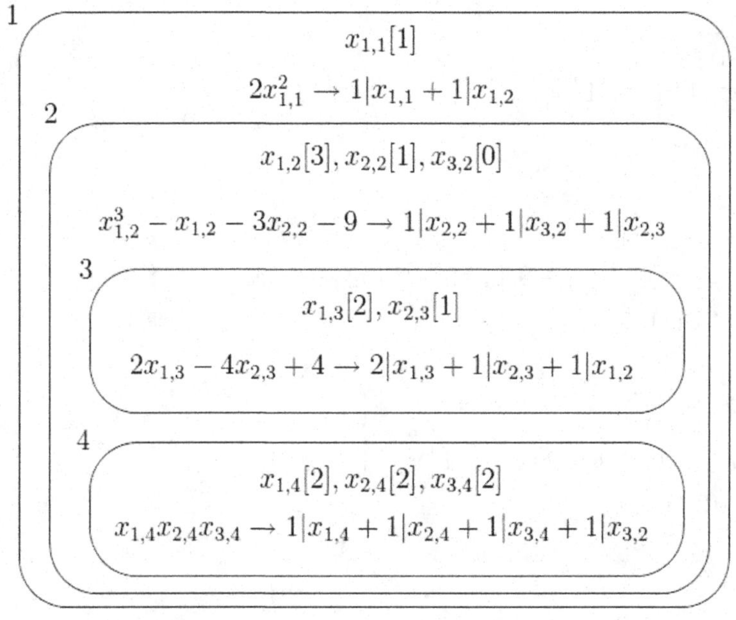

Definition 3: Enzymatic Numerical P System

An enzymatic numerical P system is defined as a construct:

$$\Pi = (m, H, \mu, (Var_1, \ E_1, Pr_1, \ Var_1(0),$$

$$Var_2, \ E_2, Pr_2, \ Var_2(0), \ldots, Var_m, \ E_m, Pr_m, \ Var_m(0))) \tag{5}$$

where E_i is a set of enzyme variables in the compartment i. The programs can have the standard, non-enzymatic form, as above, and can also have a special, enzymatic, form as in $Pr_{l,i} = (F_{l,i}(x_{1,i}, \ldots, x_{k_i,i}), e_{j,i}, c_{l,1}|v_1 + c_{l,2}|v_2 + \ldots + c_{l,n_i}|v_{n_i})$, where $e_{j,i} \in E_i$.

Example 3: A Simple Enzymatic Numerical P System

In Figure 5, a simple structure of an enzymatic numerical P system is given. For a detailed presentation of this example, the reader is referred to (Pavel & Buiu, 2011).

The condition for a production function to be active is to have the concentration of the enzyme involved in that production rule at least equal to the minimum value of the variables involved in that production function. For example, in compartment 1 from Example 3, the condition is fulfilled for the first and second program:

Figure 5. An example enzymatic numerical P system with three membranes and multiple production functions per compartment. The production functions in compartments 1 and 2 are catalyzed by enzymes e_{11}, e_{21}, and e_{12}, respectively. In compartment 3 there is a single standard production function and no enzyme

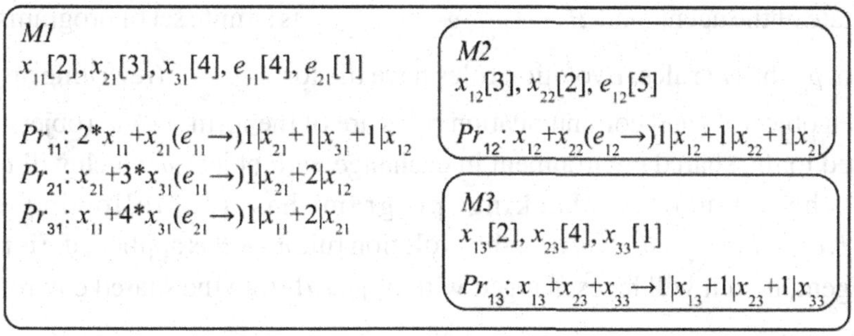

```
M1
x₁₁[2], x₂₁[3], x₃₁[4], e₁₁[4], e₂₁[1]

Pr₁₁: 2*x₁₁+x₂₁(e₁₁→)1|x₂₁+1|x₃₁+1|x₁₂
Pr₂₁: x₂₁+3*x₃₁(e₁₁→)1|x₂₁+2|x₁₂
Pr₃₁: x₁₁+4*x₃₁(e₂₁→)1|x₁₁+2|x₂₁
```

```
M2
x₁₂[3], x₂₂[2], e₁₂[5]

Pr₁₂: x₁₂+x₂₂(e₁₂→)1|x₁₂+1|x₂₂+1|x₂₁
```

```
M3
x₁₃[2], x₂₃[4], x₃₃[1]

Pr₁₃: x₁₃+x₂₃+x₃₃→1|x₁₃+1|x₂₃+1|x₃₃
```

$e_{11} > \min\left(x_{11}, x_{21}\right)$ and $e_{11} > \min\left(x_{21}, x_{31}\right)$, respectively. The third production function is not active at the first time step.

Standard and enzymatic numerical P systems can be modelled by using dedicated simulators developed at the Politehnica University of Bucharest. These are *SNUPS*, a Java simulator and *PeP*, an open-source Python simulator. Both will be briefly described in the Chapter 3 of this book.

P Colonies and P Swarms

P colonies have been introduced in (Kelemen, Kelemenová, & Păun, 2004) and have been inspired by membrane systems and colonies of formal grammars introduced in (Kelemen & Kelemenová, 1992). A P colony is a multi-agent system, in which each agent is represented by a collection of objects embedded in a single membrane. Each agent has the same number of objects inside, and the environment that surrounds the agents contains unlimited copies of a basic object e and limited copies of other objects such as a final object f. There are programs associated with agents, and each program consists of rules. The number of agents in the colony is called the degree of the colony, the height of the colony is the maximal number of programs per agent, and the capacity of the colony is the number of objects per agent (which is identical to the number of rules per program).

Definition 4: P Colony.

A P colony of capacity m is a construct:

$$\Pi = (A, e, f, B_1, ..., B_n) \tag{6}$$

where A is an alphabet (a set of objects), $e \in A$ is the basic object of the colony, $f \in A$ is the final object of the colony, and $B_i (i = \overline{1, n})$ are agents. Each agent B_i has the form (O_i, P_i), where O_i is a multiset of m copies of the basic object e (the initial state of the agent) and $P_i = \{p_{i,1}, p_{i,2}, ..., p_{i,k_i}\}$ is a finite set of programs. Each program $p_{i,j}$ has m rules. Evolution rules have the form $a \rightarrow b$ (rewriting an object a into an object b) and communication rules are of the form $c \leftrightarrow d$ (object c will be passed to the shared environment in exchange with object d which will now go inside the agent). A checking program has the following form: $< a \rightarrow b; c \leftrightarrow d \, / \, c' \leftrightarrow d' >$. After the evolution rule $a \rightarrow b$ is applied, if c is present in the agent, then it will be exchanged with object d from the shared environment;

if c is not present in the agent or d is not present in the environment, the exchange $c' \leftrightarrow d'$ will be attempted.

At the start of the computation performed by a P colony, the special basic objects e exists in the environment and in each agent. At each computation step, each agent checks whether it can apply any of its programs. If there is more than one applicable program, the agent non-deterministically chooses which program to apply. This is the parallel functioning mode. In the sequential mode, one agent is given the control at each step and applies its programs. Then, the control is passed to the next agent and so on.

In both cases, when no agent in Π is able to fire any of its programs, the computation ends and the final result is the number of final objects f in the environment.

Example 4: Increment P Colony

A simple increment P colony with one agent and two programs can be defined as:

$$\Pi_1 = (\{l_+\}, e, f, (\{e, e\}, < e \rightarrow f; e \leftrightarrow l_+ >, < l_+ \rightarrow e; f \leftrightarrow e >)) \tag{7}$$

If initially there are n objects f in the environment, at the computation halt there will be $n + 1 f$ objects (Kelemen & Kelemenová, 2005).

Initially, there are n occurrences of the final object f in the environment, and after the execution of the two programs, there will be $n + 1$ final symbols f in the environment (see Figure 6).

Example 5: Decrement P Colony

Similarly to (7), a decrement P colony can be defined as:

Figure 6. The execution of a simple addition P colony with one agent and two programs

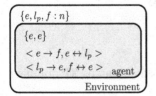
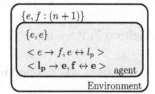

$$\Pi_2 = (\{l_{-1}, l_p, l_z\}, e, f, (\{e, e\}, <e \to e; e \leftrightarrow l_- >,$$
$$<l_- \to l_p; e \leftrightarrow f / e \leftrightarrow e >, <f \to e; l_p \leftrightarrow e >, \tag{8}$$
$$<l_p \to l_z; e \leftrightarrow e >, <e \to e; l_z \leftrightarrow e >))$$

This P colony has five programs. In the initial state, there are n occurrences of f in the environment together with one object l. The computation will end with $n - 1$ copies of object f, with l_p if there were any f in the beginning and with l_z otherwise.

The exteroceptive communication rule was introduced in (Florea & Buiu, 2016) in order to model stigmergic mechanisms needed by robots in a swarm to communicate indirectly via the environment.

Definition 5: Exteroceptive Communication Rules.

An exteroceptive communication rule extends the classical P colony communication rule (which can be considered as a proprioceptive rule) and is a communication rule of the form $c \leftrightarrow d$, which checks the presence of an object d in the shared (global) environment. If d is present, the object c will be exchanged with object d from the global environment. An exteroceptive checking program is of the following form $<a \to b; c \leftrightarrow d / c' \leftrightarrow d' >$ and will allow the object c' from the agent to be exchanged with object d' from the global environment if object c is not present in the agent or object d is not present in the global environment.

Definition 6: XP Colony.

An XP colony Λ :

$$\Lambda = (A, e, f, B_1, ..., B_n) \tag{9}$$

is a P colony of the form (6) which, in addition to the usual P colony rules, utilizes exteroceptive communication rules in order to indirectly (via the global shared environment) communicate with other XP colonies. An XP colony with exteroceptive checking rules will be called an XP colony with priority.

Definition 7: P Swarm

A P swarm was defined in (Florea & Buiu, 2016) as a colony of XP colonies:

$$\Psi = (A, E, F, \Lambda_1, \ldots, \Lambda_n) \tag{10}$$

where A is an alphabet (a set of objects), E is a multiset of basic objects of the P swarm, F is a multiset of final objects of the P swarm, and $\Lambda_i (i = \overline{1, n})$ are XP colonies of the form (9), that is, $\Lambda_i = (A_i, e_i, f_i, B_{i,1}, \ldots, B_{i,n})$.

Example 6: A P Swarm With Synchronized Increment and Decrement XP Colonies

In this example we show a practical scenario of an XP colony-to-XP colony communication using the P swarm concept (10). The increment (7) and decrement (8) P colonies, were adapted to use exteroceptive rules in order to start the increase of f objects by the increment XP colony only after the decrement XP colony has finished. Details regarding the execution processes are presented in a step by step manner in Table 1.

This P swarm is defined as follows:

$$\Psi = (\{l_+\}, e, f, \Lambda_1, \Lambda_2)$$
$$\Lambda_1 = (\{l_+, W\}, e, f, (\{e, e\}, < e \rightarrow f; e \leftrightarrow l_+ >, < l_+ \rightarrow e; f \leftrightarrow e >), (\{e, e\},$$
$$< e \leftrightarrow W; e \Leftrightarrow l_+ / e \rightarrow e >, < W \rightarrow e; l_+ \leftrightarrow e >, < W \leftrightarrow e; e \rightarrow e >))$$
$$\Lambda_2 = (\{l_{-1}, l_+, l_z\}, e, f, (\{e, e\}, < e \rightarrow e; e \leftrightarrow l_- >, < l_- \rightarrow l_+; e \leftrightarrow f / e \leftrightarrow e >,$$
$$< f \rightarrow e; l_+ \Leftrightarrow e >, < l_+ \rightarrow l_z; e \leftrightarrow e >, < e \rightarrow e; l_z \leftrightarrow e >)).$$

$$\tag{11}$$

Assuming that initially there are three f objects in the environment of each XP colony and that Λ_1 additionally contains one W object in its environment, after the execution of the P swarm is finished, the environments of Λ_1 and Λ_2 will contain 4 and 2 f objects respectively. The W object is used by the second agent of Λ_1 in a continuous exchange with the XP colony environment to prevent the XP colony from ending because otherwise no agents would be executable until Λ_2 finishes. The l_+ object is used as a signal that is exchanged through the P swarm global environment.

P colonies can compute whatever is algorithmically computable (Kelemen et al., 2004). Further interesting results on the computing power are presented in the literature, together with a number of variants of P colonies. While P colonies were

Table 1. Step by step description of the P swarm execution process: Only the executed part of conditional programs is shown. Grayed cells correspond to a P colony that has ended execution.

Step no.	Program executed		Pswarm Env	Λ_1			Λ_2	
	Λ_1	Λ_2		Env	Agent 1	Agent 2	Env	Agent 1
0				$f:3$ $W:1$	$e:2$	$e:2$	$f:3$ $l_\_:1$	$e:2$
1	$Ag2:$ $<e \leftrightarrow W, e \rightarrow e>$	$Ag1:$ $<e \leftrightarrow l_\_, e \rightarrow e>$		$f:3$	$e:2$	$e:1$ $W:1$	$f:3$	$e:1$ $l_\_:1$
2	$Ag2:$ $<W \leftrightarrow e, e \rightarrow e>$	$Ag1:$ $<l_\_ \rightarrow l_p, e \leftrightarrow f>$		$f:3$ $W:1$	$e:2$	$e:2$	$f:2$	$f:1$ $l_p:1$
3	$Ag2:$ $<e \leftrightarrow W, e \rightarrow e>$	$Ag1:$ $<f \rightarrow e, l_p \Leftrightarrow e>$	$l_p:1$ $l_p:1$	$f:3$	$e:2$	$e:1$ $W:1$	$f:2$	$e:2$
4	$Ag2:$ $<W \leftrightarrow e, e \rightarrow e>$		$l_p:1$ $l_p:1$	$f:3$ $W:1$	$e:2$	$e:2$		
5	$Ag2:$ $<e \leftrightarrow W, e \Leftrightarrow l_p>$			$f:3$	$e:2$	$W:1$ $l_p:1$		
6	$Ag2:$ $<W \rightarrow e, l_p \leftrightarrow e>$			$f:3$ $l_p:1$	$e:2$	$e:2$		
7	$Ag1:$ $<e \rightarrow f, e \leftrightarrow l_p>$			$f:3$	$f:1$ $l_p:1$	$e:2$		
8	$<l_p \rightarrow e, f \leftrightarrow e>$			$f:4$	$e:2$	$e:2$		

designed to have a passive environment, Eco-P colonies are extensions of P colonies with dynamical evolution of environments (Ludêk Cienciala & Ciencialová, 2009). P colony automata combine properties of finite automata and P colonies (Ludek Cienciala, Ciencialová, Csuhaj-Varjú, & Vazsil, 2010).

Applications of Membrane Computing

Membrane computing has attracted a large number of theoretical computer scientists and mathematicians, but the interest was also manifested in applications in a wide range of areas, including modelling biological systems, systems biology, cryptography, computer graphics, sorting/ranking, parallel architectures, evolutionary computing, economics, modelling and simulating circuits, control, robotics, etc. For a detailed introduction to the basic models of membrane systems, to the analysis of their computing power and efficiency, to applications and implementations, and research lines, the interested reader is referred to (Păun et al., 2010).

P SYSTEMS IN ROBOTICS: AN INTEGRATED POINT OF VIEW

The starting point of the use of membrane systems in the design of robot controllers was to utilize standard numerical P systems for modelling robot controllers, e.g. for avoiding obstacles or wall following. The so-called membrane controllers were in a first instance implementations of various simple control laws using standard numerical P systems (Buiu et al., 2012). Then, the enzymatic numerical P systems were introduced in order to model robot controllers (Pavel & Buiu, 2011) and new tasks have been approached using these new models, e.g. robot localization. As we mentioned, the P colonies are inspired by the way in which living organisms interact in a shared environment. An application of P colony automata for robot control is presented in (Langer, Cienciala, Ciencialová, Perdek, & Kelemenová, 2013) with additional experiments reported in (Luděk Cienciala, Ciencialová, Langer, & Perdek, 2014).

An important step further was to use the tools and power of membrane computing for multi-robot control. *Lulu* was designed as a powerful P colony and P swarm simulator allowing one to use it in interaction with robot simulators and real-world robots. In the context of swarm robotics where (possibly and desirably heterogeneous) robots have to communicate with each other directly and indirectly (via the external environment), an extension of a P colony was introduced such that the robot (which is modelled as a P colony) can check the presence of certain objects in the shared environment via its internal programs. This extension is the XP colony defined above (9).

There is a natural similarity between membrane systems and swarm robotic systems, which was first described in (Buiu & Gansari, 2014). We extend this

Table 2. Significant analogies between P systems and their variants and robotic systems (from single robots to multi-robot systems)

Property	P systems (and Their Biological Counterpart, a Living Cell, a Living Tissue)	Robotic Systems (Single or Multiple Robots)
Architecture and functions: an overview	A P system is a system of biomembranes which divide the internal cell space into discrete compartments to segregate processes and components, are selectively permeable, and play a key role in organizing complex reaction sequences, energy conservation and cell-to-cell communication.	A robotic swarm is a system of many simple robots grouped into neighborhoods and sub-swarms which divide the swarm as a whole to segregate behaviors, tasks and robots, which should be permeable to self robots but secure to non-self robots and plays a key role in power saving and swarm-swarm or swarm-human communication.
Architectural structure	P systems are composed of multiple communicating membranes that are arranged in various hierarchical structures and can be executed in parallel. The connectivity and structure of a P system is allowed to change.	Robots are composed of various perception, effector and computation devices that are interconnected. Similarly, groups of robots are composed of homogeneous or heterogeneous groups that can change in structure or role.
Modularity and hierarchy	P systems are modular and have a hierarchical structure	Robot architectures are typically modular and most of the times also hierarchical. A three-tier architecture has a behavioral control level, then up in the hierarchy is the executive layer, and the superior level is the task-planning tier.
Input-output relationships	A P system can be considered a computing device	A robot controller can be seen as implementing an input-output relationship between inputs (sensor data) and outputs (commands to actuators)

continued on next page

Table 2. Continued

Property	P systems (and Their Biological Counterpart, a Living Cell, a Living Tissue)	Robotic Systems (Single or Multiple Robots)
Interaction between agents in a shared environment	Computing agents that are modelled using P systems can interact directly with similar agents by modelling a communication device as an internal membrane that can intermediate the information exchange. An alternative option is to create an external (to the agent) membrane where messages from the entire group of agents would be placed. Both alternatives are discussed in Chapter 4.	Robots in a swarm interact directly only locally and indirectly through the environment (stigmergic mechanisms)
Interaction between agent components	Membranes in a P system can interact by exchanging information either directly, by specifying target variables using a repartition protocol (numerical P systems) or indirectly, by emitting symbolic objects with special meanings that are interpreted by the target component.	The computing device on a robot constantly reads input device values for later processing. Input devices can also independently send notifications by triggering an interrupt on the computing device. After taking a new decision, corresponding commands are sent to the output devices
Security	The biological immune system is able to discern between self and non-self cells by using the antigens.	Robotic systems must be able to make a distinction between self robots and non-self robots (possibly damaging robots). Robots must be able to join other trustworthy robots in an attempt to form groups or coalitions in order to achieve more complex tasks in a resilient way. Non-self robots must be identified and neglected (if not expelled) from interacting with self robots.
Agency and emergent intelligence	P swarms are based on simple P colonies, placed in a common environment, and their functioning leads to non-trivial emergent behaviors.	Robot swarms are based on many simple robots which interact locally and this leads to a global, emergent behavior.
Parallelism	P systems are inherently parallel and distributed computing models and some software implementations (*Lulu*, see Chapter 3) actually require the modeler to design the system in a way that avoids any conflicts between agents as they are executed in parallel.	In multi-robot coordination systems there is parallelism among subtasks and also among devices in order to increase execution efficiency.
Interaction	Interaction between regions in a P systems can be explicit or implicit.	Robots in an intentional cooperative system have knowledge and interact explicitly with other robots in the environment. Robots in a collective swarm interact mostly in an implicit way.

31

analysis in Table 2 where we summarize several points of view to the analogy between membrane systems and robots. These points can be considered as starting points for building membrane computing based controllers for single and multiple robots. A series of experiments of using P colonies and P swarms to control single and multiple interacting robots will be described in Chapter 4.

CONCLUSION

In this chapter we have introduced the basic P systems models that are relevant from the point of view of robotic applications. After describing the fundamental P system model, we described the numerical P system, the enzymatic numerical P system, the P colony, the XP colony, and the P swarm. We concluded this chapter with a detailed analogy between membrane systems and swarm robotic systems. This led us to the introduction of membrane controllers (Buiu et al., 2012) and is the foundation of our further developments presented in Chapter 4.

REFERENCES

Adleman, L. (1994). Molecular computation of solutions to combinatorial problems. *Science, 266*(5187), 1021–1024. doi:10.1126/science.7973651 PMID:7973651

Amos, M. (2010). *Theoretical and Experimental DNA Computation*. Springer.

Amos, M., Păun, G., Rozenberg, G., & Salomaa, A. (2002). Topics in the theory of DNA computing. *Theoretical Computer Science, 287*(1), 3–38. doi:10.1016/S0304-3975(02)00134-2

Buiu, C., & Gansari, M. (2014). A new model for interactions between robots in a swarm. In *Electronics, Computers and Artificial Intelligence (ECAI), 2014 6th International Conference on* (pp. 5–10). doi:10.1109/ECAI.2014.7090202

Buiu, C., Vasile, C., & Arsene, O. (2012). Development of membrane controllers for mobile robots. *Information Sciences, 187*, 33–51. doi:10.1016/j.ins.2011.10.007

Cienciala, L., & Ciencialová, L. (2009). Eco-P Colonies. In *10th Workshop on Membrane Computing WMC10, RGNC REPORT 3/2009* (pp. 201–209).

Cienciala, L., Ciencialová, L., Csuhaj-Varjú, E., & Vazsil, G. (2010). PCol Automata : Recognizing Strings with P Colonies. *8th Brainstorming Week on Membrane Computing, 1*, 65–76.

Cienciala, L., Ciencialová, L., Langer, M., & Perdek, M. (2014). The Abilities of P Colony Based Models in Robot Control. In *Proceedings of the 15th International Conference on Membrane Computing* (pp. 155–168). Springer International Publishing. doi:10.1007/978-3-319-14370-5_11

Florea, A. G., & Buiu, C. (2016). Development of a software simulator for P colonies. Applications in robotics. *International Journal of Unconventional Computing, 12*(2–3), 189–205.

Head, T. (1987). Formal language theory and DNA: An analysis of the generative capacity of specific recombinant behaviors. *Bulletin of Mathematical Biology, 49*(6), 737–759. doi:10.1007/BF02481771 PMID:2832024

Holland, J. H. (1992). *Adaptation in natural and artificial systems : an introductory analysis with applications to biology, control, and artificial intelligence*. MIT Press.

Kelemen, J., & Kelemenová, A. (1992). A grammar-theoretic treatment of multiagent systems. *Cybernetics and Systems, 23*(6), 621–633. doi:10.1080/01969729208927485

Kelemen, J., & Kelemenová, A. (2005). On P colonies, a simple bio-chemically inspired model of computation. In I. Rudas (Ed.), *6th International Symposium on Computational Intelligence* (pp. 40–56). Budapest Tech.

Kelemen, J., Kelemenová, A., & Păun, G. (2004). Preview of P colonies: A biochemically inspired computing model. In *Workshop and Tutorial Proceedings, Ninth International Conference on the Simulation and Synthesis of Living Systems(AlifeIX), (M. Bedau et al., eds.)* (pp. 82–86). Boston, Mass.

Langer, M., Cienciala, L., Ciencialová, L., Perdek, M., & Kelemenová, A. (2013). *An Application of the PCol Automata in Robot Control. In 11th Brainstorming Week on Membrane Computing* (pp. 153–164).

Nguyen, V., Kearney, D., & Gioiosa, G. (2010). A region-oriented hardware implementation for membrane computing applications. Lecture Notes in Computer Science (Including Subseries Lecture Notes in Artificial Intelligence and Lecture Notes in Bioinformatics), 5957 LNCS, 385–409. http://doi.org/ doi:<ALIGNMENT. qj></ALIGNMENT>10.1007/978-3-642-11467-0_27

Păun, G. (2000). Computing with Membranes. *Journal of Computer and System Sciences, 61*(1), 108–143. doi:10.1006/jcss.1999.1693

Păun, G., & Păun, R. (2006, January 1). Membrane Computing and Economics: Numerical P Systems. *Fundamenta Informaticae*. IOS Press.

Păun, G., Rozenberg, G., & Salomaa, A. (2010). *The Oxford Handbook of Membrane Computing*. Oxford University Press, Inc. doi:10.1007/978-3-642-11467-0

Pavel, A. B., Arsene, O., & Buiu, C. (2010). Enzymatic numerical P systems - a new class of membrane computing systems. In *2010 IEEE Fifth International Conference on Bio-Inspired Computing: Theories and Applications (BIC-TA)* (pp. 1331–1336). IEEE. http://doi.org/ doi:10.1109/BICTA.2010.5645071

Pavel, A. B., & Buiu, C. (2011). Using enzymatic numerical P systems for modeling mobile robot controllers. *Natural Computing, 11*(3), 387–393. doi:10.1007/s11047-011-9286-5

Pavel, A. B., Vasile, C. I., & Dumitrache, I. (2012). *Robot Localization Implemented with Enzymatic Numerical P Systems* (pp. 204–215). Springer Berlin Heidelberg; doi:10.1007/978-3-642-31525-1_18

Phillips, R., Kondev, J., & Theriot, J. (2012). *Physical Biology of the Cell*. Garland Science.

Reif, J. H., & H., J. (1995). Parallel molecular computation. In *Proceedings of the seventh annual ACM symposium on Parallel algorithms and architectures - SPAA '95* (pp. 213–223). ACM Press. http://doi.org/<ALIGNMENT.qj></ALIGNMENT>10.1145/215399.215446

Rothemund, P. W. K., & Winfree, E. (2000). The program-size complexity of self-assembled squares (extended abstract). In *Proceedings of the thirty-second annual ACM symposium on Theory of computing - STOC '00* (pp. 459–468). ACM Press. http://doi.org/ doi:<ALIGNMENT.qj></ALIGNMENT>10.1145/335305.335358

Chapter 3
Software for Membrane Computing

ABSTRACT

In order to use membrane computing models for real life applications there is a real need for software that can read a model from some form of input media and afterwards execute it according to the execution rules that are specified in the definition of the model. Another requirement of this software application is for it to be capable of interfacing the computing model with the real world. This chapter discusses how this problem was solved along the years by various researchers around the world. After presenting notable examples from the literature, the discussion continues with a detailed presentation of three membrane computing simulators that have been developed by the authors at the Laboratory of Natural Computing and Robotics at the Politehnica University of Bucharest, Romania.

INTRODUCTION

Membrane computing has attracted a notable interest from the scientific community, both in the theoretical foundations and in the applications in multiple domains. Shortly after the introduction of P systems as a distributed and parallel computing model, P systems simulators have been developed, tested, and made available to the membrane computing community. The purpose of this chapter is to give an overview of existing membrane computing simulators and to discuss their capabilities. Included in this discussion are detailed presentations of the simulators we have designed: *SNUPS* (for simulating standard and enzymatic numerical P systems), *Lulu* (for simulating P colonies and P swarms) and *PeP* (an on-going open-source project to develop a

DOI: 10.4018/978-1-5225-2280-5.ch003

standard/enzymatic numerical P system simulator). These membrane computing simulators required tackling several design and implementation challenges that are described in *Simulator Design and Implementation Challenges*, found within this chapter. The understanding of this chapter is based upon the assimilation of the fundamental principles of membrane computing presented in Chapter 2.

MEMBRANE COMPUTING SIMULATORS

In general, the existing software for simulating membrane computing models can be divided into three categories taking into account the paradigm they use: (1) sequential (C, Java, C++, etc.), (2) software-based parallelization (of which some of the best known are *Open MPI* and *OpenMP*), and (3) hardware-based parallelization (*FPGAs*, etc.). *Graphics Processing Units* (GPUs) follow a hybrid paradigm and offer a many-core platform with high parallelism at low cost (Martínez-del-Amor, Macías-Ramos, Valencia-Cabrera, Riscos-Núñez, & Pérez-Jiménez, 2014). Our discussion and presentation will follow this categorization.

One of the earliest works in the area of simulating P systems is reported in (Ciobanu & Paraschiv, 2002) and describes the *Membrane Simulator* which provides a graphical simulation for two basic P systems (the hierarchical cell system and the active membrane system).

An ANSI C library with simple data structures that facilitated the *in silico* study of P systems was proposed in (Nicolau Jr, Solana, Fulga, & Nicolau, 2002). There can be implemented both active and non-active membranes, and actions for dissolving, dividing and creating new membranes.

(Borrego et al., 2007) continued the work dedicated to the graphical simulation of P systems and presented a tool called *Tissue Simulato*r which supports the understanding of the basic structure and functioning of tissue P systems with cell division. This tool was developed in Java (to analyze the input data) and C# (used for the graphical user interface and for the kernel of the application). This software is no longer available to the community. A software tool for assisting the formal verification of spiking neural P systems (*SNPS*) has been proposed in (Gutiérrez-Naranjo, Pérez-Jiménez, & Ramírez-Martínez, 2008).

The membrane computing paradigm lies at the basis of *Cyto-Sim* (Sedwards & Mazza, 2007) which included a formal language and also a Java stochastic simulator of membrane systems. *Cyto-Sim* allowed the use of Petri nets based models and was able to import and export SBML files and to export MATLAB files. In (Spicher, Michel, Cieslak, Giavitto, & Prusinkiewicz, 2008) there is a report on an implementation of stochastic P systems using *MGS*, a spatially explicit programming language.

Psim, a simulation tool based on metabolic algorithms, was presented and discussed in (Bianco & Castellini, 2007), allowing the simulation of metabolic P systems. *MetaPlab* is an interesting development and presents itself in the form of a virtual laboratory implemented in Java, available at http://mplab.sci.univr.it/ and allowing the understanding and simulation of the internal mechanisms of biological systems (Castellini & Manca, 2008).

SNUPS, a Java software tool for modelling and simulation of standard numerical P systems and of enzymatic numerical P systems (Arsene, Buiu, & Popescu, 2011) is presented together with a detailed example in (Buiu, Arsene, Cipu, & Patrascu, 2011). This application software allows the development of a wide range of applications from modelling and simulation of ordinary differential equations, to the design and simulation of computational blocks for cognitive architectures and of membrane controllers for autonomous mobile robots. In the case of enzymatic numerical P systems, *SNUPS* simulates membrane systems with enzyme-like variables, which allow the existence of more than one production function in each membrane and keep the deterministic nature of the system.

SNUPS can be used on any operating system as long as a Java Virtual Machine and the *Swing* graphical interface library are installed. The application was tested on Windows OS and Mac OS until now. In its interface there is highlighted the menu of *SNUPS* (see Figure 1). In the top of the figure there are three tabs: "membrane" (the default one), which shows in the left of the figure the membrane structure; "symbols", where symbols can be created or/and added to a certain membrane; "rules", where the evolution rules of the membrane will be edited. In the bottom of the figure there is the possibility to modify the default number of cycles (which is 1), and the button for starting the computation.

In Figure 2 it is detailed the menu "file", that has two options: "open", which can open a previous, saved membrane structure; and "save as", which can save the membrane structure to an XML file.

In Figure 3 it is presented the "symbols" tab. The symbols can be added to the membrane by clicking in the "select" column, and the initial value can be edited in the "initial value" column. Beside these symbols, known and used in *SNUPS* with one rule per membrane, there are another type of symbols, enzymes, which can be added to a certain membrane by clicking the "enzyme" column. As seen in the Figure 3, by clicking the "enzyme" column of a certain symbol (in order to add that enzyme symbol to a membrane), the "select" column of the same symbol it is marked automatically. *SNUPS* also allows adding other symbols, by using "create new symbol" from the bottom of the figure. After writing the symbol, it is necessary to press the "validate" button and then the "add" button. When the symbols are added, they appear in the left vertical side of the application in the membrane structure.

Figure 1. SNUPS: Main interface

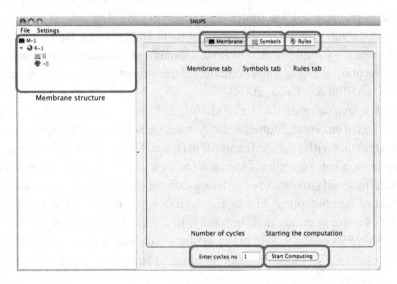

Figure 2. SNUPS interface: The File menu

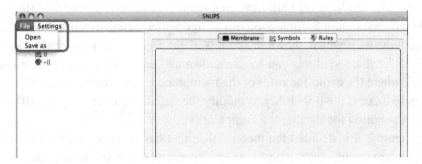

In the Figure 4 below there are highlighted the four sections of the "rules" tab (in this order):

- "Contribution table", where contribution rules can be edited (rows can be added, symbols and their contribution value can be edited). After the "apply" button is pressed, the contribution values are committed and can be seen in the left vertical side of the application, in the membrane structure.
- "Rules editor", where the production function is edited. After the "apply" button is pressed, the rules are committed and can be seen in the membrane structure.

Figure 3. SNUPS: The "symbols" tab

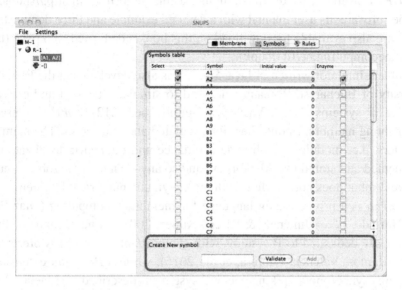

Figure 4. SNUPS: The "rules" tab

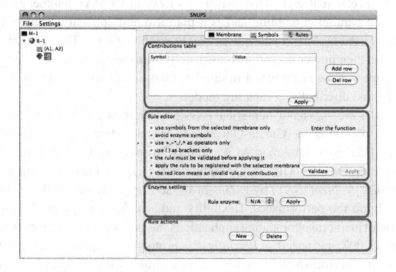

- "Enzyme setting" is the panel where an enzyme can be applied (used) in the rule edited above in the "rule editor". One enzyme can be applied to one or more production functions from the same membrane.
- "Rule action", where new rules can be created (new rules have a blank contribution table, rule editor and enzyme setting)

SNUPS is available for download in its executable form from http://snups.buiu. net. The software, its user manual with a detailed example and three demonstration movies are also available for download on this book's dedicated web page (http:// membranecomputing.net/IGIbook).

Another simulator written in Java, *SimP*, was also developed at the Politehnica University of Bucharest, Romania and used to simulate standard and enzymatic numerical P systems (Pavel, Vasile, & Dumitrache, 2012). *SimP* was used for implementing membrane controllers for simulated and real robots. The membrane controllers, e.g. modules for obstacle avoidance and odometric localization, are implemented and stored in XML format, and so any change in the robot simulator or the real robot does not interfere with the enzymatic numerical P system.

P-Lingua is a programming language for membrane computing (Díaz-Pernil, Pérez-Hurtado, Pérez-Jiménez, & Riscos-Núñez, 2009). A new simulator for the *pLinguaCore* dedicated to P systems with symport/antiport rules is presented in (Macías-Ramos, Valencia-Cabrera, et al., 2015), while a *P-Lingua* extension for simulating asynchronous spiking neural P systems is described in (Macías-Ramos, Pérez-Jiménez, Song, & Pan, 2015).

Computing with GPUs (Graphics Processing Units) has been shown to be an attractive and efficient approach to data parallelization. CUDA is another prominent approach in parallel computing and is a platform and application programming interface created by *Nvidia* which uses CUDA-enabled GPUs. The simulation of P systems and their numerous variants can benefit from the use of parallel computing, and this appears to be a natural way to simulate them as they were originally designed as parallel and distributed computing models.

A parallelized enzymatic numerical P systems based of graphic processing units (GPUs) is presented in (García-Quismondo, Macías-Ramos, & Pérez-Jiménez, 2013). In (Martínez-del-Amor, M.A., Pérez-Carrasco, J., Pérez-Jiménez, 2013) there is presented a new GPU-based simulator for a quadratic time solution to the *Satisfiability problem* (SAT) by means of tissue P systems with cell division. An early study on the performance, flexibility and scalability of parallel computing platforms for membrane computing applications was performed in (Nguyen, Kearney, & Gioiosa, 2007) and tested on *Reconfig-P*, a prototype computing platform. GPUs were shown to perform better than CPUs for the simulation of P systems with active membrane (Cecilia et al., 2010).

The computational power of GPUs was shown to benefit the simulation of *Population Dynamics P systems* (Martínez-del-Amor, Pérez-Hurtado, Gastalver-Rubio, Elster, & Pérez-Jiménez, 2012). A recent survey (Martínez-del-Amor et al., 2015) reviews the existing parallel P simulators using GPUs, also providing guidelines for future developments of parallel implementations and simulations of P systems.

LULU: AN OPEN-SOURCE SIMULATOR
FOR P COLONIES AND P SWARMS

Lulu is an open-source Python implementation of a P colony and P swarm simulator that was introduced in (Florea & Buiu, 2016a) and is available on Github (Florea & Buiu, 2016b). Among the features of this simulator are:

1. Hardware and software portability;
2. Reading input P colonies/P swarms from text files written using a syntax that closely resembles that used for the mathematical definitions;
3. Simulation of a P colony/P swarm with:
    a. Deterministic and stochastic program selection;
    b. Evolution, communication, exteroceptive and checking rules;
4. Printing a structured view of intermediate and final states of the model;
5. Interfacing with other applications.

The P colonies or P swarms that are to be executed by *Lulu* are read from text files and stored into internal structures. An example input file for the increment P colony (see Chapter 2, Example 4) is presented alongside the mathematical definition in Figure 5. The decrement P colony that was also presented theoretically in Chapter 2 (Example 5) is described using the *Lulu* input language alongside a P swarm (Chapter 2, Example 6) that allows the formation of a chain reaction between the increment and decrement P colonies also in Figure 5.

Among notable differences that are present in the *Lulu* input file are the naming of agents, the use of composite items (such as *pi*) that are defined using curly braces and semicolons that are placed at the end of instructions. The denomination of agents was done to help study the execution of the P colony while the syntax differences were needed to allow the automatic parsing of the input file into object oriented structures. These internal structures are presented using an UML Class diagram in Figure 6.

The attributes of each P swarm component have been identified based on the specific details that were needed by the simulation algorithm in order to decide whether a program is executable or not and consequently execute it. For instance, the *Rule* class can retain all of the details of even the most complex conditional rules (that are actually two rules in one), all in a generic manner. Examples of such details (class members) are: left/right hand side symbolic objects (*lhs/rhs*), *main_type* used to store the rule type in case of simple rules or to mark the rules as conditional. All of these attributes are duplicated (using the *alt_* prefix) to accommodate conditional rules. Programs are a simple list of rules while agents maintain a list of programs and a link to their associated P colony. The *Counter* type that is used for storing

Figure 5. The increment P colony described using the mathematical definition (a) and the input file format used by Lulu (b). The decrement P colony (c) and a P swarm (d) that integrates the increment and decrement XP colonies into a chain of operations (decrement and afterwards increment)

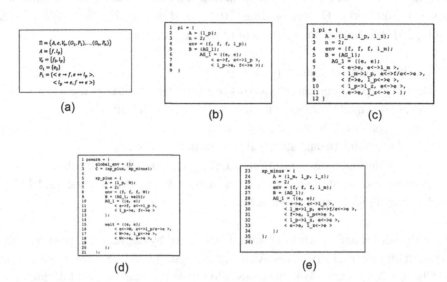

(a)

(b)

(c)

(d)

(e)

multisets is part of the *collections* modules that is included in the standard Python distribution. This class offers specific methods for storing and interacting with multisets and is used for agent multisets, P colony environment and P swarm global environment. More details regarding the implementation and internal functioning of the simulator as well as a complete user guide and documentation can be found on http://membranecomputing.net/IGIbook/.

After reading the input file and creating a structured internal representation of the P colony or P swarm, the simulator can begin the execution of simulation steps. The simulator was designed to follow the maximally parallel execution strategy that is used (most often) for P colonies and P swarms. Under this strategy, each agent that has an executable program at one simulation step must execute it and if there are multiple such programs then choose one non-deterministically and execute it. Also, if there are multiple executable agents at one time, they must all execute their programs in parallel. The first requirement was implemented in *Lulu* while the second one that refers to the parallel execution of agents was partially implemented and this option will now be explained and also described graphically in Figure 7.

The P colony, as well as the majority of membrane computing models (including the P system), has been designed as a parallel computing model, where agents are executed in parallel. In terms of software design, this design consideration can be

Figure 6. UML Class diagram of Lulu

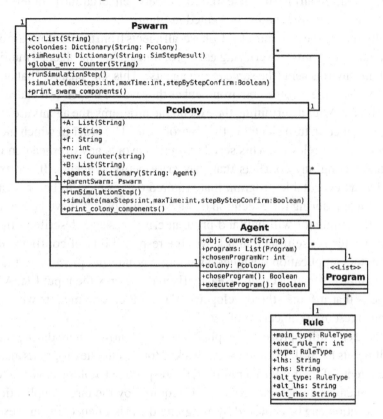

implemented using multiple concurrent threads, a concept that is generally used to simultaneously execute multiple code blocks and has wide support across many platforms. The benefit of using threads to execute agents, one thread for each agent, is that the theoretical execution parallelism is achieved. The downside is that agents may need to exchange objects with the P colony environment and can only do so in an exclusive manner because otherwise the contents of the multiset could get corrupted because of multiple simultaneous changes. The mutual exclusion needed to ensure that only one thread modifies the P colony environment at any given time has the consequence of blocking any other concurrent threads that are waiting to do the same operation. For simple operations, in terms of computing time, such as the ones required to modify the contents of a multiset according to the rules of a program, the increased overhead of managing threads and ensuring exclusive access to common multisets can diminish the theoretical speedup that could be obtained, sometimes up to the point of increasing the execution time.

For these reasons an intermediate solution, in between serial and parallel execution paradigms, has been used and is presented in Figure 7.

Initially, during the program check phase, all agents from the P colony are checked for executable programs by verifying each program against the agent's multiset and the P colony environment (the same for all agents). This step ensures that all agents initially see the same P colony environment as they would if each was executed on a separate thread. After determining the executable programs, the agents are executed in a serial manner, in turn changing the environment. The order in which the agents are executed is not relevant. This serial execution avoids any overhead and is also useful in detecting any conflicts that may arise between agents. If for example, *Agent 2* has an executable program that required one *f* object to be present in the environment and during the execution of *Agent 1*, all *f* objects have been removed/replaced, the simulator will halt and print an error message describing the cause of the error. This is useful in that it leaves the responsibility of conflict avoidance to the P colony application designer, avoiding any internal processing that could potentially hide errors that could be easily fixed using only the input file. A second advantage is that it forces the developer to design P colony models where agents do not reach resource (object) conflict.

The same execution process is applied also to P swarms where during the check phase, all agents from all P colonies are checked for executable programs against the current contents of their internal multiset, the respective P colony environment and the P swarm environment. Nonetheless, if required by the target application, true parallel execution can be achieved by assigning the task of checking and executing each P colony agent to a thread.

An example output of the simulator using its text-based interface, for the execution of the increment P colony, is shown in Figure 8. The simulator presents the user with information regarding the currently executed program and the current state of the multisets that form the P colony or P swarm.

Figure 7. P colony execution process in Lulu

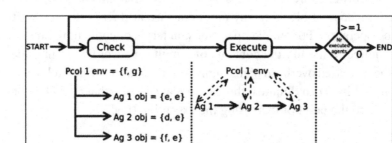

From the early design phases, *Lulu* was intended as a P colony interpreter that could be integrated into more complex applications. For this reason, both the Python and C versions of the simulator are libraries that can be easily included in other projects. The internal functioning of the simulator has also been designed in a modular way in order to allow external control of the simulation process, as seen in Figure 9. The input data can be processed before each simulation step and if required, a numeric to symbolic value conversion can take place and insert or replace symbolic objects in the P colony with ones that correspond to the processed input values. Similarly, after the execution process for the current step has finished, a symbolic to numeric conversion can take place and then send numeric values to output devices. Complex applications can be developed using this pattern and examples of single and multi-robot controllers are presented in Chapter 4.

The *Kilobot* robot is a simple low-cost oriented robot that features vibrating motors for movement and an RGB LED that can represent 64 colors and in terms of input devices, the robot features an ambient light sensor and an infrared communication device that allows it to detect the distance from neighboring robots (Rubenstein, Ahler, & Nagpal, 2012). The *Kilobot* is described with more details in Chapter 1 of this book.

Based on the previously discussed generic robot controller, a basic controller for the *Kilobot* robot that allowed the control of the robot's movement directions was proposed in (Florea & Buiu, 2016a). An extended controller that offers control over

Figure 8. Lulu output after executing the increment P colony

```
$ python sim.py input_ag_increment.txt        Pcolony = {
                                                   env = [('f', 3), ('e', 1)]
INFO      Reading input file                      AG_1 = {
INFO      building Pcolony                             obj = [('1_p', 1), ('f', 1)]
                                               }
Pcolony = {                                    INFO      Starting simulation step 1
    A = ['1_p']                                INFO      Agent AG_1 is runnable
    n = 2                                      INFO      1 runnable agents
    env = [('f', 3), ('1_p', 1), ('e', 1)]     INFO      Running Agent AG_1  P1 = < 1_p -> e,
    B = ['AG_1']                               f <-> e >
        AG_1 = {                               INFO          Simulation  step  finished
            obj = [('e', 2)]                   successfully
            programs = {                       Pcolony = {
                P0 = <                             env = [('f', 4), ('e', 1)]
                    e -> f                         AG_1 = {
                    e <-> 1_p                          obj = [('e', 2)]
                >                              }
                P1 = <                         INFO      Starting simulation step 2
                    1_p -> e                   INFO      0 runnable agents
                    f <-> e                    Pcolony = {
                >                                  env = [('f', 4), ('e', 1)]
            }                                      AG_1 = {
        }                                              obj = [('e', 2)]
}                                              }
                                               INFO          Simulation finished successfully
INFO      Starting simulation step 0           after 2 steps and 0.001134 seconds; End state
INFO      Agent AG_1 is runnable               below:
INFO      1 runnable agents                    Pcolony = {
INFO      Running Agent AG_1  P0 = < e -> f, e     env = [('f', 4), ('e', 1)]
<-> 1_p >                                          AG_1 = {
INFO          Simulation  step  finished               obj = [('e', 2)]
```

45

Figure 9. Lulu P colony execution loop and integration within a generic controller

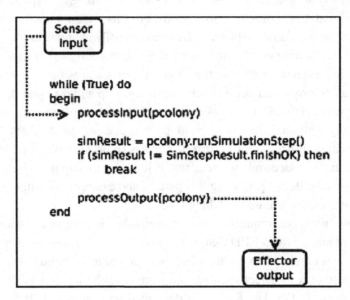

all input and output devices present on the *Kilobot* robot (presented in Chapter 1) was introduced in (Florea & Buiu, 2016) and is presented in Figure 10.

The arrows that connect the controller input and output to the corresponding agents are dotted (as in Figure 9) because of the background processing that is required to convert numeric values to symbolic values and vice versa. For output devices this process is straightforward because of the nature of the vibrating movement system present on the *Kilobot* that allows only three possible directions: left, right, forward.

Figure 10. The current Kilobot controller implemented using P colonies

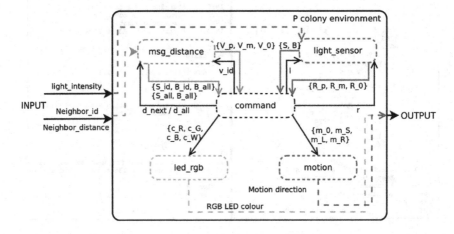

The same is valid for the RGB LED where each color can be directly mapped to a symbolic object due to the limited set of values. On the other hand, in the case of input devices, distance or light intensity measurements are received as integer values and must be discretized into meaningful symbolic values. A short description of all symbolic objects that were used in this controller is presented in Table 1.

Portability is an important requirement of current open-source projects and has represented a design objective throughout the implementation process. Portability refers here not only to the ability to execute the same code on various machines (and operating systems) but also to the simplicity of the porting task. The simulator was written entirely in Python 3 without any external (operating system specific) dependencies, allowing it to be executed on any machine that can run the Python 3 interpreter, without requiring the user to search for a particular version of a library or Python 3 interpreter. Another important fact in this direction is that *Lulu* was designed without any Graphical User Interface (GUI), relying on a text-based terminal interface. This option on one hand reduces the dependency list, memory consumption and overall system requirements and on the other hand increases execution speed. Recently, a new implementation of the simulator in the C language (C99 standard) was written and tested. This compiled version has the same features as the Python version and is intended for use on CPU architectures that do not have support for the Python interpreter, such as embedded systems. The C implementation of *Lulu*

Table 1. Short description of the basic objects used for symbolic conversion of input data.

Object	Creating Agent	Meaning
d_id	Command	Request the current distance from robot *id*
d_all	Command	Request the minimum distance from my current neighbors
S_id, B_id	Msg_distance	Distance from robot *id* is Small/Big
S_all	Msg_distance	There is at least one neighbor robot that is at distance Small from the robot executing the P colony
B_all	Msg_distance	All neighbor robots are at distance Big from the robot executing the P colony
v_id	Command	Request the distance variation for robot *id*
V_p, V_m, V_0	Msg_distance	The distance from robot *id* has increased (p), decreased (m), remained constant (0)
L	Command	Request the current ambient light level
S, B	Light_sensor	The current ambient light is Small/Big
r	Command	Request the ambient light variation
R_p, R_m, R_0	Light_sensor	The ambient light value has increased (p), decreased (m), remained constant (0)

allows for the integration with a dedicated *Kilobot* simulator (also written in C) entitled *Kilombo* that enables large (on the order of thousands) numbers of *Kilobots* to be simulated at speeds up to 100 times the real speed using the same code that can be executed by real robots without any modification (just a recompilation) (Jansson et al., 2015).

Due to the fact that the same code has to be executable on both a PC (with Gigabytes of memory) and a micro-controller (with 32 Kilobytes of memory), careful consideration was taken during the porting process to avoid using all of the available memory. The general conversion procedure is depicted in Figure 11, where all of the steps required to process an input file and transform it into C code are presented.

All of the rules that the *Lulu* simulator currently supports are written by *Lulu* in the *rules* library for use in the C application.

The object oriented structure obtained after processing the input P colony is used by the *Lulu_c* Python script to write a structure representation of the P colony into the *instance* library. This set of structures follows closely the object oriented structure but replaces all memory consuming data structures that were used in Python such as dictionaries and vectors with simple arrays and accompanying functions that interpret the arrays in order to store and operate on more complex data structures such as multisets.

Figure 11. The processing steps and applications involved in transforming a human readable P swarm definition into a P colony that can be simulated locally on a Kilobot (real or simulated using Kilombo)

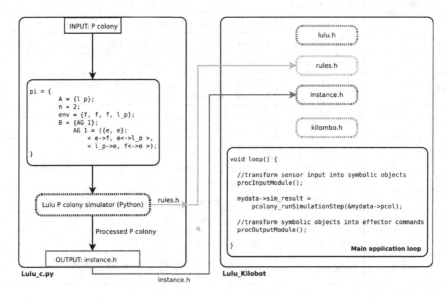

A comparison between these two implementations of *Lulu* as well as the robot simulation platforms available to each has been performed using the following hardware and software configuration: an Intel i3-4000M CPU with a memory size of 8GB DDR3 on a Linux 64 bit operating system.

A direct performance comparison between the two implementations of the *Lulu* P colony simulator has been performed by means of comparing the CPU time required and peak memory consumption during 20 trials of executing a modified version of the decrement P colony. The goal of the test decrement P colony was to ultimately remove all of the *f* objects in the environment, not only one, this way increasing the computational time so that it could be recorded more easily. The modified P colony was defined as:

$$\Pi = (\{l_-, \ l_p, \ l_z\}, \ e, f, (\{e, e\}, <e \rightarrow e; e \leftrightarrow l_- >, <l_- \rightarrow l_p; e \leftrightarrow f / e \leftrightarrow e >,$$
$$<f \rightarrow f; l_p \rightarrow l_- >, <f \rightarrow e; l_- \leftrightarrow e >, <l_p \rightarrow l_z; e \leftrightarrow e >, <e \rightarrow e; l_z \leftrightarrow e >)). \tag{1}$$

The modified decrement P colony (1) was executed by each *Lulu* implementation (C and Python) and the CPU time that was needed to execute the P colony (until no more steps were executable) was recorded using the *perf* application, distributed alongside the Linux kernel. This profiling application is capable of recording and processing performance information related to the CPU from both software events generated by the Linux kernel as well as hardware events generated by the processor itself, without affecting kernel or process performance (de Melo, 2010). The application was used to record several performance parameters regarding the execution of the P colony on each of the two simulator variants, using the *perf stat* command and recording the *task-clock* parameter in milliseconds. The peak memory usage level was recorded using the *Massif* memory profiler, part of the *Valgrind* suite of debugging and profiling tools (Seward, Nethercote, & Weidendorfer, 2008). These profiling tools were preferred over (language) internal performance analysis methods in order to use the same information gathering tool for two very different language types (compiled versus interpreted).

The numerical results for the two types of tests are presented in Table 2. While both the CPU time and peak memory usage values are substantially smaller for the C implementation, one other difference is that of the variability of measurements. This variability, especially for the CPU time, can be noticed using the *standard deviation* parameter but also using the box and whiskers plot presented in Figure 12. Due to the steep difference in the scale of the values (the mean value of the CPU time for the Python implementation is 58.1103 times bigger than the one written in C), the values were processed by means of centering around the mean (the mean was subtracted from all values) and afterwards scaled (all values were divided by the standard deviation), all performed using the *scale* function from the *R* statistical

Table 2. Comparison between the implementations of the Lulu simulator in C and Python respectively

Criteria	CPU Time (Millisecond)		Peak Memory Usage (MB)	
	Lulu C	Lulu Python	Lulu C	Lulu Python
Maximum	1.794608	95.37479	6.258	41.31
Minimum	1.461098	85.64236	6.258	41.29
Mean	1.533835	89.13163	6.258	41.3
Standard deviation	0.08074655	2.596766	0	0.003244428

language (R Core Team, 2016). After the scale operation has been performed, an analysis of the plot confirms the previous observation that the variability of the CPU time for the Python implementation is higher than that of the C variant, because of the difference in height between the two plots (including whiskers). The higher variability and CPU time for the Python implementation is not expected to grow substantially for more complex input files because this CPU time includes the time required for the interpretation of the Python source code that is afterwards executed. Although an important fact in this comparison, this interpretation time is expected to remain largely constant because the source code remains the same.

In what follows, a comparison of the two controller implementations, in C and Python, that interface to *Kilombo* and *V-REP* respectively is performed, by executing the dispersion algorithm on both controllers. Details about the algorithm as well as the P colony used to describe it are provided in Chapter 4.

The runtime performance is the first criteria considered and because *Kilombo* is written in a compiled language (C) and *V-REP* in an interpreted language (Lua), the time penalty is in the order of milliseconds and grows as the Lua source code grows. Another important aspect in this sense is that the robots that were simulated in *V-REP* were controlled remotely through a dedicated communication protocol defined by the simulator that introduced a severe time penalty and limited the swarm size to ~20 robots, depending on the CPU used. *Kilombo* simulations were transparently executing the same program on each robot and this proved to be a much more scalable approach, allowing the simulation of up to 1000 robots.

The overall comparison between the previous P colony based *Kilobot* controller that used *V-REP* and the current one that uses *Kilombo* is presented in Table 3. In each cell, the + sign denotes the cell that refers to a better value for the respective criteria.

Figure 12. Boxplot of the (scaled) CPU times required to process a 30 objects decrement P colony by each of the two implementations of Lulu. The mean value of the scaled CPU times is zero.

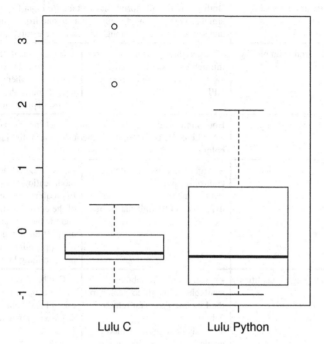

Criteria 6) and 8) refer to tests that were performed on an Intel i3-4000M CPU with a memory size of 8GB DDR3 on a Linux 64 bit operating system.

Criteria 8), *Last robot activation interval for 20 robots*, refers to the time interval needed for the last robot to receive a new signal. This value is relevant because in both cases robots are processed in a loop, starting with robot 0 and ending with robot 19.

PEP: AN OPEN-SOURCE SIMULATOR FOR SIMULATING NUMERICAL P SYSTEMS AND THEIR VARIANTS

PeP is similar in scope and purpose to *Lulu* but allows the use of direct numerical variables, without requiring additional numeric-symbolic conversions. A direct consequence of this fact is that numerical models, such as a *Proportional–Integral–Derivative* (PID) controller for example, can be directly modelled using membrane computing concepts and organization. Although *SNUPS*, the previously introduced numerical P systems simulator was already capable of executing this type of systems, it was based on Java and this limited its application domain to computers that could

Table 3. An overview of advantages and disadvantages of Lulu implementations

No.	Comparison criteria	Lulu Python + V-REP	Lulu C + Kilombo
1	Program readability	Both simulating platforms interpret a Lulu input file and during simulation they convert numeric data to and from symbolic values and so the P colony is enough to describe the entire program.	
2	Program complexity	- Higher, due to the manual loop through each robot from the swarm and the remote control API.	+ Lower, due to the transparent switch from one robot's control loop to the other and the integration of the simulator in the user application.
3	Reusability	Both simulating platforms use *Lulu* as backend for executing the robot behavior therefore they inherit the reusability provided by P colonies with specialized agents	
4	Flexibility	+ Higher, because the remote-control script is written in Python and benefits from the dynamic and flexible nature of Python.	- Lower, in part because every modification of the *Lulu* input file requires the regeneration of the corresponding C code and the recompilation of the application and also because this application uses C as programming language.
5	Debugging facilities	- No standard Lua debugger available and no direct access to the Lua interpreter for remote debugging. The only debugging method available are user-added prints to the console	+ Can be directly controlled by the debugger, allowing the use of breakpoints, step-by-step execution and many other features.
6	Runtime performance (maximum number of robots before lag of more than one second appears in robot state switch)	- 40	+ 15000
7	Exception handling	+ Lua interpreter prints out a detailed exception message and the line number that triggered it.	+ Independent application prints only an error title but using a debugger we can have access to the exact line number that triggered the execution halt as well as the execution tree (backtrace)
8	Last robot activation interval for 20 robots	- 4.57 seconds	+ Less than 100ms

run the Java Virtual Machine. On the other hand, *PeP*, as *Lulu*, will be offered in both Python and C versions in order to be deployable to embedded or even micro-controller driven systems.

The basic structure of the simulator is adapted (as was the case of *Lulu*) to the memory requirements of processing and executing a numerical P system (with and without enzymes). This structure is depicted in Figure 13. using an UML Class diagram.

Comparing the class structure to a numerical P system one can note that the notion of hierarchical structure among membranes has been implemented in the simulator through the use of references (of the Membrane class) to its parent membrane and to a list of descendent membranes. This is necessary during execution of the numerical P systems in order to access and modify variables from neighboring membranes. Another benefit is that the system can be traversed (and possibly executed) in a recursive manner, reducing the memory footprint of the application during execution.

Programs contain a pair of production and distribution functions and as stated in the theoretical definition, one membrane can contain more than one program and for this reason, membranes maintain a list of programs. The integration with other control software and devices remains a design objective and can be achieved through integrating *PeP* (a Python module or C library) into the target application in order to reduce communication bandwidth requirements.

Figure 13. PeP UML Class structure

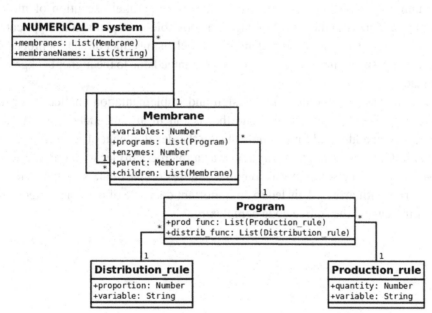

SIMULATOR DESIGN AND IMPLEMENTATION CHALLENGES

All software simulators of physical systems require models that abstract the physical execution context, removing unnecessary details. This is a compromise that is meant to allow the simulator to either handle complex scenarios or to increase the simulation speed, when compared to a real life execution. Depending on the requirements of the simulator, the level of abstraction can vary and with it, the difficulty in developing the target system model and most importantly the simulator architecture.

The membrane computing (P system and P colony) simulators that were presented above obviously do not have to model a physical system. Instead, they execute a computational model that is described in mathematical terms with respect to structure and functionality. Some abstractions are already made by the developer of the computational model (e.g., P system enzymes that do not follow the exact behavior of real enzymes) but that does not imply that transforming a mathematical computational model into an application is completely straightforward. Some of the complications that arise during the design and implementation stages of this type of simulators will be now outlined. The solutions to each issue have been presented throughout this chapter.

One of the first aspects to consider is flexibility, here in the context of computing models that continue to evolve, adding new concepts that increase the computational power. This is the main reason why an object oriented architecture has been preferred. This also allows for a simpler integration into other applications and this was a design consideration for all membrane simulators.

Another example to consider is the theoretical parallel execution of most P system (and derivate) models. Although it is possible to implement an application that executes each membrane (or agent) in parallel, using either threads or processes, the execution speed increase would not be guaranteed due to the higher background overhead.

A small memory footprint is a design and implementation challenge for any software application, particularly one that is ran in an embedded context. This topic has been addressed for example during the development of *Lulu* (the variant written in C) because the target system was a microcontroller with only 32 Kb of RAM memory. The solution was to replace more complex data structures such as dictionaries with arrays. This led to an important decrease of memory usage at the cost of increasing the source code complexity.

CONCLUSION

A review of membrane computing simulators has been presented in this chapter. The simulators were presented in chronological order. This section also included detailed presentations of the simulators that have been implemented by the authors at the *Laboratory of Natural Computing and Robotics*.

Several design and implementation challenges of developing a membrane computing simulator have been outlined in this chapter. These challenges were paired with the solutions used by the authors in an attempt to provide a clear perspective over the relationship between the theoretical model and the practical software and hardware restrictions.

REFERENCES

Arsene, O., Buiu, C., & Popescu, N. (2011). SNUPS - A Simulator For Numerical Membrane Computing. *International Journal of Innovative Computing*, *7*(6), 3509–3522.

Bianco, L., & Castellini, A. (2007). Psim: A Computational Platform for Metabolic P Systems. In Membrane Computing (pp. 1–20). Berlin, Heidelberg: Springer Berlin Heidelberg. http://doi.org/ doi:10.1007/978-3-540-77312-2_1

Borrego, R., Ropero, –, Díaz, D., Pernil, –, Pérez, M. J., & Jiménez, –. (2007). Tissue Simulator: A Graphical Tool for Tissue P Systems. In *Proceedings of the International Workshop, Automata for Cellular and Molecular Computing* (pp. 23–34). Budapest.

Buiu, C., Arsene, O., Cipu, C., & Patrascu, M. (2011). A software tool for modeling and simulation of numerical P systems. *Bio Systems*, *103*(3), 442–447. doi:10.1016/j.biosystems.2010.11.013 PMID:21146581

Castellini, A., & Manca, V. (2008). MetaPlab: A Computational Framework for Metabolic P Systems. In Membrane Computing (pp. 157–168). Berlin, Heidelberg: Springer Berlin Heidelberg. http://doi.org/ doi:10.1007/978-3-540-95885-7_12

Cecilia, J. M., García, J. M., Guerrero, G. D., Martínez-del-Amor, M. A., Pérez-Hurtado, I., & Pérez-Jiménez, M. J. (2010). Simulation of P systems with active membranes on CUDA. *Briefings in Bioinformatics*, *11*(3), 313–322. doi:10.1093/bib/bbp064 PMID:20038568

Ciobanu, G., & Paraschiv, D. (2002). P System Software Simulator. *Fundamenta Informaticae*, *49*(1–3), 61–66.

de Melo, A. C. (2010). The new linux'perf'tools. In Slides from Linux Kongress (Vol. 18).

Díaz-Pernil, D., Pérez-Hurtado, I., Pérez-Jiménez, M. J., & Riscos-Núñez, A. (2009). A P-Lingua Programming Environment for Membrane Computing. In D. W. Corne, P. Frisco, G. Păun, G. Rozenberg, & A. Salomaa (Eds.), *Membrane Computing* (Vol. 5391, pp. 187–203). Berlin, Heidelberg: Springer Berlin Heidelberg. doi:10.1007/978-3-540-95885-7_14

Florea, A. G., & Buiu, C. (2016). Development of a software simulator for P colonies. Applications in robotics. *International Journal of Unconventional Computing*, *12*(2–3), 189–205.

Florea, A. G., & Buiu, C. (2016). Lulu - an open-source software simulator of P colonies and P swarms. Retrieved from https://github.com/andrei91ro/lulu_pcol_sim

Florea, A. G., & Buiu, C. (2016). Synchronized dispersion of robotic swarms using XP colonies. In *Electronics, Computers and Artificial Intelligence (ECAI), 2016 8th Edition International Conference on*.

García-Quismondo, M., Macías-Ramos, L. F., & Pérez-Jiménez, M. J. (2013). *Implementing Enzymatic Numerical P Systems for AI Applications by Means of Graphic Processing Units* (pp. 137–159). Springer Berlin Heidelberg; doi:10.1007/978-3-642-34422-0_10

Gutiérrez-Naranjo, M. A., Pérez-Jiménez, M. J., & Ramírez-Martínez, D. (2008). A software tool for verification of Spiking Neural P Systems. *Natural Computing*, *7*(4), 485–497. doi:10.1007/s11047-008-9083-y

Jansson, F., Hartley, M., Hinsch, M., Slavkov, I., Carranza, N., & Olsson, T. S. G. … Grieneisen, V. A. (2015). Kilombo: a Kilobot simulator to enable effective research in swarm robotics. *arXiv Preprint arXiv:1511.04285*.

Macías-Ramos, L. F., Pérez-Jiménez, M. J., Song, T., & Pan, L. (2015). Extending Simulation of Asynchronous Spiking Neural P Systems in P–Lingua. *Fundamenta Informaticae*, *136*(3), 253–267.

Macías-Ramos, L. F., Valencia-Cabrera, L., Song, B., Song, T., Pan, L., & Pérez-Jiménez, M. J. (2015). A P_Lingua Based Simulator for P Systems with Symport/Antiport Rules. *Fundamenta Informaticae*, *139*(2), 211–227. doi:10.3233/FI-2015-1232

Martínez-del-Amor, M. A., García-Quismondo, M., Macías-Ramos, L. F., Valencia-Cabrera, L., Riscos-Núñez, A., & Pérez-Jiménez, M. J. (2015). Simulating P systems on GPU devices: A survey. *Fundamenta Informaticae*, *136*(3), 269–284. doi:10.3233/FI-2015-1157

Martínez-del-Amor, M. A., Macías-Ramos, L. F., Valencia-Cabrera, L., Riscos-Núñez, A., & Pérez-Jiménez, M. J. (2014). *Accelerated Simulation of P Systems on the GPU: A Survey* (pp. 308–312). Springer Berlin Heidelberg; doi:10.1007/978-3-662-45049-9_50

Martínez-del-Amor, M. A., Pérez-Carrasco, J., & Pérez-Jiménez, M. J. (2013). Characterizing the parallel simulation of P systems on the GPU. *International Journal of Unconventional Computing*, *9*(5–6), 405–424.

Martínez-del-Amor, M. A., Pérez-Hurtado, I., Gastalver-Rubio, A., Elster, A. C., & Pérez-Jiménez, M. J. (2012). *Population Dynamics P Systems on CUDA* (pp. 247–266). Springer Berlin Heidelberg; doi:10.1007/978-3-642-33636-2_15

Nguyen, V., Kearney, D., & Gioiosa, G. (2007). Balancing Performance, Flexibility, and Scalability in a Parallel Computing Platform for Membrane Computing Applications. In *Membrane Computing* (pp. 385–413). Berlin, Heidelberg: Springer Berlin Heidelberg; doi:10.1007/978-3-540-77312-2_24

Nicolau, D. V. Jr, Solana, G., Fulga, F., & Nicolau, D. V. (2002). A C Library for Simulating P Systems. *Fundamenta Informaticae*, *49*(1–3), 241–248.

Pavel, A. B., Vasile, C. I., & Dumitrache, I. (2012). *Robot Localization Implemented with Enzymatic Numerical P Systems* (pp. 204–215). Springer Berlin Heidelberg; doi:10.1007/978-3-642-31525-1_18

R Core Team. (2016). R: A Language and Environment for Statistical Computing. Retrieved from https://www.r-project.org/

Rubenstein, M., Ahler, C., & Nagpal, R. (2012). Kilobot: A low cost scalable robot system for collective behaviors. In *2012 IEEE International Conference on Robotics and Automation (ICRA)* (pp. 3293–3298). Ieee. http://doi.org/ doi:10.1109/ ICRA.2012.6224638

Sedwards, S., & Mazza, T. (2007). Cyto-Sim: A formal language model and stochastic simulator of membrane-enclosed biochemical processes. *Bioinformatics (Oxford, England), 23*(20), 2800–2802. doi:10.1093/bioinformatics/btm416 PMID:17855418

Seward, J., Nethercote, N., & Weidendorfer, J. (2008). *Valgrind 3.3-Advanced Debugging and Profiling for GNU/Linux applications*. Network Theory Ltd.

Spicher, A., Michel, O., Cieslak, M., Giavitto, J.-L., & Prusinkiewicz, P. (2008). Stochastic P systems and the simulation of biochemical processes with dynamic compartments. *Bio Systems, 91*(3), 458–472. doi:10.1016/j.biosystems.2006.12.009 PMID:17728055

Chapter 4
P Colonies for the Control of Single and Multiple Robots

ABSTRACT

The fourth chapter is dedicated to the topic of controlling single and multiple robots using P colonies. The open-source P colony simulator Lulu is used as a software basis that allowed for the development of a discrete robot controller. The chapter is organized around a set of experiments that are ordered by difficulty and are described using the P colony model and execution diagrams. These diagrams allow the reader to view when and what will be the response of the robot to a given input signal. Several screenshots are also presented to aid in understanding this process. All of the experiments can be fully replicated by using the input files provided within the chapter or the on-line guidelines that are available at http://membranecomputing. net/IGIBook.

INTRODUCTION

Robots are machines capable of executing complex actions automatically and are composed of various types of modules (sub-systems). By analogy, as we detailed in Chapter 2, a P colony is also composed of several interacting agents and so modelling a robot using a P colony can be generally seen as creating a mapping between robot modules and P colony agents. Due to the fact that P colonies function using symbolic objects, there is a natural need for bi-directional conversion methods between symbolic values used by the P colony and numerical ones used by the sensors and effectors of the robot.

DOI: 10.4018/978-1-5225-2280-5.ch004

From the different types of robots that have been developed so far, the best candidates for implementing a P colony controller are robots whose input data and output commands can be easily discretized into symbolic values.

Swarm robotics was defined in (Şahin, 2005) as "the study of how large numbers of relatively simple physically embodied agents can be designed such that a desired collective behavior emerges from the local interactions among agents and between the agents and the environment". Several defining characteristics have also been pointed out (Şahin, 2005):

1. Autonomy;
2. Large number of robots;
3. Few homogeneous groups of robots;
4. Relatively incapable or inefficient robots;
5. Robots with local sensing and communication.

P swarms have been proposed as symbolic modeling tools for robotic swarms by grouping multiple P colonies, each associated with an individual robot (Buiu & Gansari, 2014).

Various control algorithms that have been implemented using P colonies and P swarms are presented in this chapter. Each experiment is described using the source P colony that is written in the language accepted by *Lulu* (our P colony and P swarm simulator which was described in detail in Chapter 3) and can be viewed online at the dedicated webpage, alongside input files, P colony state diagrams such as the one presented in Figure 1 and instructions for replicating the experiments using open-source applications.

Figure 1. P colony state diagram

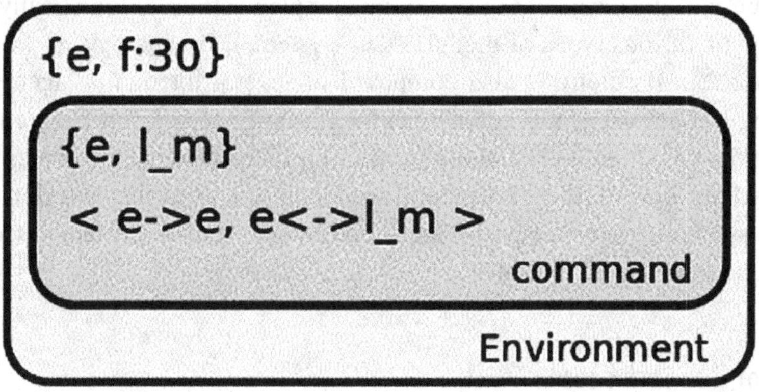

P colony state diagrams are used throughout this chapter to describe the effects of executing a given program, in terms of environment and agent multiset content changes. Only the agents that had an executable program for that simulation step are included within the environment membrane. In the example presented in Figure 1, the environment multiset contains the elementary object *e* (that is never depleted according to the P colony definition from Chapter 2) and 30 *f* objects while the *command* agent contains one elementary object *e* and one *l_m* object, after the execution of the < *e->e, e<->l_m* > program.

The expected behavior of the robot or swarm is presented graphically using robot state diagrams such as the one in Figure 2. Although the *Kilobots* do not have any direct means of determining their global or relative orientation, the global orientation is still presented using the black triangle on top (in this case) of the robot for the purpose of visualizing the evolution of a given algorithm.

The experiments presented in this chapter are ordered by complexity, starting from single robot experiments and continuing with experiments with multiple robots, again sorted by complexity.

SINGLE ROBOT EXPERIMENTS

A series of experiments for controlling single robots are presented in this section in order to describe the basic functioning principle of the *Lulu_Kilobot* controller. Although robotic swarms are the main subject of interest of this work, single robot experiments allow for an in depth analysis of concepts and algorithmic tools that can enhance multi-robot experiments. Such enhancements include the ability to control a robot using remote (or preprogrammed) commands sent through a buffered input device (found within *Tape Rules,* Chapter 4) and the ability to react to the elapse of time (found within *Timer Interrupt*, Chapter 4).

Figure 2. Example robot state diagram:The current LED color is marked with the initial letter (R-red, G-green, B-blue) and the direction is indicated using a dashed arrow

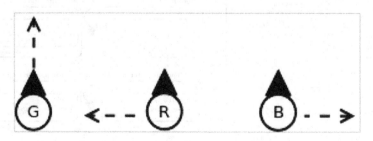

Moving a Fixed Distance in a Specified Direction

The simple task of moving a mobile robot for a fixed distance can be implemented using either feedback from motion sensors (such as wheel encoders) or from a simple decreasing counter variable. The latter was used in this case because a single *Kilobot* robot does not have any means of measuring the distance it has moved nor the current orientation. As such, the counter variable under a P colony controller was considered the multiplicity of a symbolic object that is decreased at each simulation step. When no objects of that type are available in the environment, the robot movement is stopped.

In the example presented in Figure 3, one *Kilobot* simulated using the *V-REP* simulator must follow a straight path for a fixed distance that is encoded by defining 5 *f* objects in the P colony environment (*env*), as seen in Figure 3a, line 4. The *l_m* object is used as a subtraction trigger and its presence in the environment determines agent *command* to attempt to subtract an *f* object (Figure 3b step 2). If the operation is successful, an *m_S* object is sent in the environment and at the next step received by the *motion* agent. The conversion between symbolic input/output objects and numerical commands used by the robot takes place in between execution steps (see *Lulu: An Open Source Simulator for P Colonies and P Swarms*, Chapter 3) and thus the robot starts moving straight before the simulator starts to execute the fifth simulation step. Once started, the robot motion is continuous and is not affected by subsequent *move straight (m_S)* commands. After all *f* objects have been removed

Figure 3. Lulu input file and execution diagram for moving a robot 5 "steps" forward

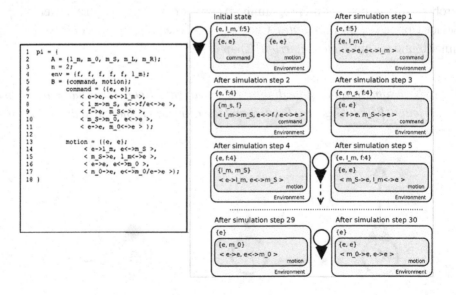

(consumed) from the environment, the *command* agent will issue a *move stop (m_0)* object that once received by the *motion* agent will determine the movement stop of the robot.

Tape Rules

A P colony algorithm for driving a robot down a straight static course was presented in the previous section. One method of creating a more complex course or of *remote* controlling a P colony robot controller is the use of *tape* rules. These special types of rules were proposed in (Cienciala, Ciencialová, Csuhaj-Varjú, & Vazsil, 2010) as means of introducing objects (signals) into the P colony in a step-by-step manner by placing a string of symbolic objects on an input tape that is read by the P colony agents. In order to maintain the parallel execution of agents and prevent the formation of a queue of two or more agents that are waiting for objects on the tape, in the tape concept defined in (Cienciala et al., 2010) it is also specified that all agents that can read a given object from the tape will do so simultaneously and only after this step will the tape advance to the next symbolic object. The first implementation of a robot controller that used P colonies with tape rules was reported in (Langer, Cienciala, Ciencialová, Perdek, & Kelemenová, 2013).

The simultaneous control of movement direction and RGB LED color of the robot is considered a test case for this concept and presented using the robot state diagram and *Lulu* input code in Figure 4. In order to simplify the controller and also to emphasize on the idea of parallel processing of the tape by multiple agents, the *command* agent has been stripped out and the *motion* and *rgb_led* agents are responsible for reading the input tape and reacting accordingly. The only type of tape rule that was used is *tape evolution* (denoted by the *-T->* operator) that transforms the left-hand-side object in the agent into the right-hand-side object if this object is the current object at the reading end of the tape. The *tape* multiset was also defined in order to allow the definition of the initial contents of the tape. During the P colony execution, the tape is read from left-to-right until all of the objects are read but external control could be simply achieved by introducing objects from the keyboard or other input device or even from communication devices.

The initial state of the robot is stationary with no color turned on. As soon as the first object is read from the tape, each agent responds in a specific way, by either generating a move or color command that is executed before the next simulation step begins. After executing all agents that had an executable program for the object read from the tape, the P colony continues with a next step where a new symbolic object is read, and so on until all objects are read from the tape or a halting condition is reached, a P colony configuration in which no agents have executable programs and therefore the execution ends. Similarly to the previous experiment, the tape

Figure 4. Lulu input file and robot state diagram for moving a robot on a pre-determined course that involves three directions and corresponding colors

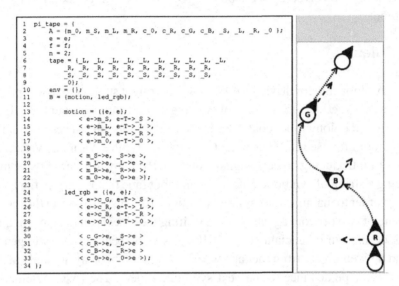

```
1  pi_tape = {
2     A = {m_0, m_S, m_L, m_R, c_0, c_R, c_G, c_B, _S, _L, _R, _0 };
3     e = e;
4     f = f;
5     n = 2;
6     tape = { _L, _L, _L, _L, _L, _L, _L, _L, _L, _L,
7              _R, _R, _R, _R, _R, _R, _R, _R, _R, _R,
8              _S, _S, _S, _S, _S, _S, _S, _S, _S,
9              _0};
10    env = {};
11    B = {motion, led_rgb};
12
13       motion = {{e, e};
14          < e->m_S, e-T->_S >,
15          < e->m_L, e-T->_L >,
16          < e->m_R, e-T->_R >,
17          < e->m_0, e-T->_0 >,
18
19          < m_S->e, _S->e >,
20          < m_L->e, _L->e >,
21          < m_R->e, _R->e >,
22          < m_0->e, _0->e >};
23
24       led_rgb = {{e, e};
25          < e->c_G, e-T->_S >,
26          < e->c_R, e-T->_L >,
27          < e->c_B, e-T->_R >,
28          < e->c_0, e-T->_0 >,
29
30          < c_G->e, _S->e >
31          < c_R->e, _L->e >
32          < c_B->e, _R->e >
33          < c_0->e, _0->e >};
34 };
```

multiset contains multiple objects of the same type in order to extend the period during which a robot takes a specific action.

Timer Interrupt

Although simple subtraction P colonies or the use of tape rules can provide a method of setting the robot on a static course of various levels of complexity, these methods cannot guarantee precision in the sense of multiple runs of the controller yielding the same path with small enough variations. The different paths that may be recorded are caused by the different execution times of the P colony and these cannot be accurately predicted especially due to the nature of multi-task computing systems, even though smaller variances can still be noticed in single-task systems.

In cases where high precision/repeatability is a requirement, a solution can be the use of an external (to the simulator) trigger such as a timer interrupt. The timer device was modelled as a P colony agent using the same request/response pattern that was used for the other input/output devices present on the *Kilobot*, by defining request objects t_start/t_stop for activating/deactivating the timer and the timer interrupt object *T_I* that is emitted by the *timer* agent into the P colony environment after a pre-specified time interval has passed.

Considering the previous course that was used for the demonstration of the *tape rule* concept we now re-implement this task using the *timer* agent. The algorithm is presented in Figure 5 using the *Lulu* input language and in Figure 6 using a robot and P colony state diagram. The definition of the *motion* and *led_rgb* agents was not included as it is the same as that presented in Figure 3a. For readability purposes, the P colony state diagram is described only until the first output command is issued.

In the case of the *tape* rule implementation, the succession of directions (and colors) was defined by the order of the symbolic objects on tape. For the timer based implementation, the current state had to be remembered from one execution step to another and also taken into consideration when deciding the next action. In this sense, the model does resemble a Finite State Machine (FSM), by retaining the current state within the agent's multiset and executing state transitions under certain conditions (such as the trigger of a timer interrupt). By comparison, an FSM is defined using states, transitions (between states) and actions, reacting to changes in the environment that are marked as events (Dash, Dasgupta, Chepada, & Halder, 2011) (a timer interrupt in this case). Events can trigger state transitions if the correct conditions are met and actions are taken before entering, after leaving or while in a state (Dash et al., 2011). The main difference between a P colony algorithm and an FSM one is that in the case of the latter, only the decision part of the algorithm is modelled while the rest of the interplay between algorithm components is written

Figure 5. Lulu input file for moving a robot on a static course using a timer interrupt to trigger direction and color changes

```
1   pi_static = {
2       A = {1_m, m_0, m_S, m_L, m_R, c_R, c_G, c_B, c_W,
3            ST0, ST1, ST2, ST3, t_start, t_stop, T_ON, T_I};
4       n = 4;
5       env = {};
6       B = {command, motion, led_rgb, timer};
7       command = ({t_start, e, e, e};
8           < t_start<->e, e->ST0, e->e, e->e >,
9
10          # go to ST1 left_red
11          < e<->T_I, e->c_R, e->m_L, ST0->ST0 >,
12          < T_I->e, c_R<->e, m_L<->e, ST0->ST1 >,
13
14          # go to ST2 right_blue
15          < e<->T_I, e->c_B, e->m_R, ST1->ST1 >,
16          < T_I->e, c_B<->e, m_R<->e, ST1->ST2 >,
17
18          # go to ST3 straight_green
19          < e<->T_I, e->c_G, e->m_S, ST2->ST2 >,
20          < T_I->e, c_G<->e, m_S<->e, ST2->ST3 >
21
22          # go to final stop state
23          < e<->T_I, e->c_0, e->m_0, ST3->t_stop >,
24          < T_I->e, c_0<->e, m_0<->e, t_stop<->e >);
25
26      timer = ({e, e, e, e};
27          < e->e, e->e, e->e, e<->t_start >,
28          # receive timer stop object no matter if the timer is on or not
29          < T_ON->e / e->e, e->e, e->e, e<->t_stop >,
30          # send a timer interrupt or keep waiting
31          < T_ON->T_ON, T_I<->e / e->e, e->e, e->e >);
32  };
```

Figure 6. P colony and robot state diagrams for the static course algorithm implemented using a timer interrupt

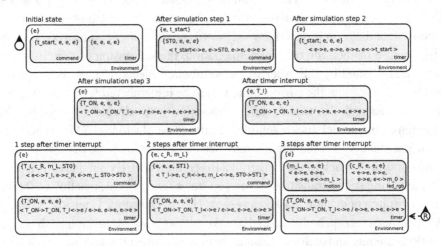

directly in the programming language used for the target application whereas in the case P colonies, a greater extent of the algorithm logic is modelled.

Another constraint that is imposed by the internal functioning of the simulator (according to the P colony definition) is that a P colony continues execution as long as there is at least one agent that has an executable program. For this reason, the *timer* agent was designed such that once started (upon receiving the *t_start* object from the environment, see Figure 5, line 27) will continue to execute the program on line 31 until it is deactivated using a *t_off* object. The same program (line 31) is also responsible for publishing the timer interrupt object (*T_I*) that is generated in the agent's multiset at regular intervals by the simulator backend.

The entire execution set of steps that take place from the start of the simulation up until the processing of the first timer interrupt is presented in Figure 6, where one can notice that at certain simulation steps there are multiple agents that are executed simultaneously, using the inherent parallelism upon which the P colony computing model was created.

MULTI-ROBOT EXPERIMENTS

In this chapter we extend the context of the previous experiments by simulating multiple robots at first independent and afterwards interacting with a precise purpose.

Figure 7. Lulu_Kilobot configuration file (a) and expected robot state diagram (b) for the control of 5 robots using different P colonies

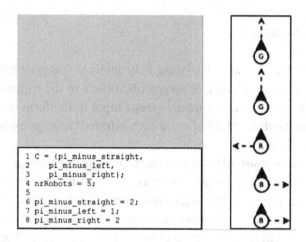

Multiple Independent Robots

The first multi-robot experiment that is to be presented is that of five independent robots that are controlled by three different types of P colonies: two robots move straight, one left and two right, as seen in Figure 5. The source code of the first P colony is the same as in Figure 3a for the robot that moves straight forward while the other two robots are controlled using modified P colonies where the move straight *m_S* object was replaced with move left (*m_L*) and move right (*m_R*).

In order to accomplish this task, one can pursue two different paths, depending on the implementation details: programming each P colony separately on each robot/group in the case of a distributed control architecture or by defining a mapping between one or more robots and the corresponding P colony type when using a centralized control system. The latter is used in this example and the robot – P colony association is specified using a separate configuration file shown in Figure 7a.

The configuration file is read by the *Lulu_Kilobot* controller after reading a P swarm input file that contains the definitions of all three P colonies. The P colony list, defined in Figure 7a on the first three lines, is used to enumerate all of the P colonies (from the P swarm input file) that are to be used in the simulation. The attributions on lines 6-8 directly specify how many robots will be assigned to a given P colony. The robot-to-P colony assignments take place in the same order as that of the P colony list in Figure 7a lines 1-3 so that the first (from top to down) two robots are assigned the *pi_minus_straight* P colony and so on. At each simulation step, the P swarm execution loop passes through all defined P colonies and executes one simulation step for each P colony.

Collision Avoidance

Up until this point, all controllers used only internal (proprioceptive) feedback such as the presence or absence of a symbolic object or the triggering of a timer interrupt. In this experiment we model sensor input in the form of robot to robot distance measurements that take place at each infrared message exchange between two *Kilobot* robots.

The test scenario consists of two robots that face each other and that are programmed to move forward, using the simple subtraction controller presented in Figure 3. During the move, the *command* agent will continuously request a measurement of the distance to the other robot and as soon as this distance falls below a given threshold value, it will stop moving in order to avoid a collision. The simulation is performed using two independent P colonies that are executed together under the P swarm context (see Chapter 2).

An extract of one of the two P colonies is presented in Figure 8 alongside three screenshots taken during simulation. The controller of each robot requests the distance from the other robot, using the d_0 object in this case. The *msg_distance* agent is responsible for responding to such a request object by replacing it before the next simulation step begins with one of the distance measurement objects, B_0 or S_0 that correspond to distance *big* or distance *small* from the specified robot, by comparing the numerical distance with a static threshold. This binary set of values was chosen because of the high measurement variation that was observed for real *Kilobot* robots, mostly due to lighting conditions and imperfections on the running surface. An increased number of discretization steps can be implemented for more precise measurement needs or sensing devices.

Synchronization Between a Pair of Robots Using the P Swarm

Interactions between members of a swarm represent a key element in the description of a swarm of robots and can happen either directly or indirectly. The latter can be described as stigmergy, an interaction method where the traces left in the environment by other swarm members influence the next decision (Heylighen, 2016). An example of indirect interaction was presented in the previous subsection where two robots were capable of sensing the distance from one another and prevent a collision.

A direct interaction experiment is presented in this section where the concept of P swarm is used to model the message exchange between two robots. Each robot is controlled by an individual P colony and the message exchange takes place between

Figure 8. Definition (a) of the control and sensor input agents, command and msg_distance respectively as part of the collision avoidance algorithm and pictures of the two simulated robots (b) taken during execution

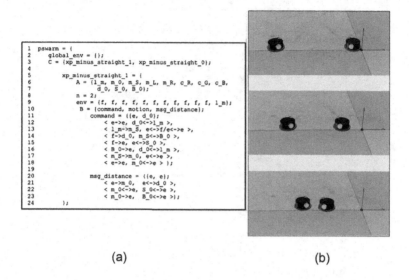

```
1   pswarm = {
2     global_env = {};
3     C = {xp_minus_straight_1, xp_minus_straight_0};
4
5     xp_minus_straight_1 = {
6       A = {l_m, m_0, m_S, m_L, m_R, c_R, c_G, c_B,
7            d_0, S_0, B_0};
8       n = 2;
9       env = {f, f, f, f, f, f, f, f, f, f, l_m};
10      B = {command, motion, msg_distance};
11        command = {{e, d_0};
12          < e->e,  d_0<->l_m >,
13          < l_m->m_S,  e<->f/e<->e >,
14          < f->d_0,  m_S<->B_0 >,
15          < f->e,  e<->S_0 >,
16          < B_0->e,  d_0<->l_m >,
17          < m_S->m_0,  e<->e >,
18          < e->e,  m_0<->e > };
19
20        msg_distance = {{e, e};
21          < e->m_0,  e<->d_0 >,
22          < m_0<->e,  S_0<->e >,
23          < m_0->e,  B_0<->e >};
24    };
```

(a)   (b)

agents from different P colonies using *exteroceptive* rules that are similar in functioning to the *communication* rules used by agents from the same P colony. P colonies that make use of exteroceptive rules and are part of a P swarm are known as XP colonies (described in Chapter 2). In this experiment, a synchronization task between two robots is presented where the first robot moves a certain distance forward (using a simple decrement controller) and once finished, signals the second robot to start moving by sending a *signal* object into the P swarm global environment multiset. The *Lulu* input file for this signaling algorithm is presented in Figure 9.

Both XP colonies closely resemble the basic decrement P colony in order to move the robot a fixed distance in a specified direction. The XP colony that is used by the second robot (moving left) includes an additional agent, named *heartbeat* due to its task of maintaining the XP colony executable while waiting for the first robot's XP colony to send the signal object s_L. The continuous exchange of the wait object W that is executed by this agent throughout the waiting period can be visualized using the XP colony and robot state diagram presented in Figure 10. The *heartbeat* agent halts after the signal object has been received by the *command* agent and published in the XP colony environment by reaching an internal multiset configuration (s_L, e) for which there are no executable programs.

During execution there are steps when several agents from the same XP colony as well as multiple XP colonies are executed simultaneously, such as on row 2 from Figure 10. Such situations are often encountered for P/XP colony algorithms and

Figure 9. P swarm global environment synchronization algorithm described using the Lulu input language

```
1   pswarm = {
2       global_env = {};
3       C = {xp_minus_straight, xp_minus_left};
4
5           xp_minus_straight = {
6               A = {l_m, m_0, m_S, m_L, m_R, c_R, c_G, c_B, s_L};
7               n = 2;
8               env = {f, f, f, f, f, f, f, f, f, f, f, f, f, f, l_m};
9               B = {command, motion};
10                  command = {{e, e};
11                      < e->e, e<->l_m >,
12                      < l_m->m_S, e<->f/e<->e >,
13                      < f->e, m_S<->e >,
14                      < m_S->m_0, e->s_L >,
15                      < s_L<=>e, m_0<->e > };
16          };
17
18          xp_minus_left = {
19              A = {l_m, m_0, m_S, m_L, m_R, c_R, c_G, c_B, s_L, W};
20              n = 2;
21              env = {f, f, f, f, f, f, f, f, f, f, f, f, f, W};
22              B = {command, motion, heartbeat};
23                  command = {{e, e};
24                      < e->e, e<=>s_L >,
25                      < s_L<->e, e->l_m >,
26
27                      < e->e, e<->l_m >,
28                      < l_m->m_L, e<->f/e<->e >,
29                      < f->e, m_L<->e >,
30                      < m_L->m_0, e<->e >,
31                      < e->e, m_0<->e > };
32
33                  heartbeat = {{e, e};
34                      < e->e, e<->s_L / e<->W >,
35                      < s_l->e, W->e >,
36                      < W<->e, e->e >};
37          };
38  }
```

Figure 10. P swarm and robot execution diagram for the P swarm global environment signal algorithm

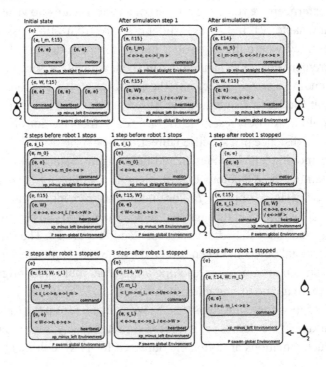

require attention to the expected symbolic object interchange that will take place, mainly to anticipate and avoid potential conflicts between agents.

Dispersion

In the swarm robotics definition, it is stated that large numbers of individuals interact locally for the swarm to achieve a desired (emergent) behavior (Şahin, 2005). The dispersion of a swarm of robots is presented in this experiment with the task for each robot to move beyond a specified distance away from any other robots that are in its communication range. A requirement of the algorithm is to allow robots to adapt to a dynamic neighborhood and to be scalable in terms of the number of robots in the swarm.

The distance perception agent *msg_distance* is used in this experiment with input data that is similar to that used for the collision avoidance experiment. The request/response objects have a slightly different meaning in order to support the scalability and dynamic neighborhood requirements. Whereas the initial *msg_distance* agent responded to requests regarding the distance to a specific robot (for example *d_2* was used to request the distance from robot with id 2), in this experiment it responds to requests regarding all of the nearby (within communication range) robots, using the *d_all* object that has the meaning of *what is the minimum distance from all nearby robots?* The response objects have also been changed to *B_all* meaning that no robots that are in communication range are closer than the specified distance and *S_all* meaning that at least one robot is in the specified distance. The background processing is done by the controller by simply comparing distances from the current neighbors. Using this pair of request and response objects, the need for knowing (and testing) all robot ids (scalability) and also the dependence on a local set of neighbors that can change over time (dynamic neighborhood) are removed.

An extract of the control algorithm is presented in Figure 11 along with screenshots taken at various execution steps during a 10 robot simulation in *V-REP*. The capacity of the P colony is three in order to simultaneously emit a motion command as well as an RGB LED color command, to aid in visualizing the current decision of each robot.

Upon receiving a distance *small* object (*S_all*), the algorithm stochastically chooses a new movement direction (and corresponding color) in order to move away from the close-distance neighbor. The stochastic election of one program from a set of executable programs is included in the P colony definition (and supported by *Lulu*). For this reason, programs on lines 13 to 18 from Figure 11a have been designed to be all four executable at same time if an *S_all* object is present in the environment. Four programs have been defined (the program on line 14 is a duplicate of the one from line 13) in order to make the robot choose with a probability of 50% to go straight

Figure 11. Dispersion algorithm (a) and screenshots taken during different execution states of the algorithm (b), starting from the initial state (top) and ending in the dispersed state (bottom)

```
1   pi_disperse = {
2       A = {1_m, m_O, m_S, m_L, m_R, c_R, c_G, c_B, c_O, c_W,
3             d_all, S_all, B_all};
4       n = 3;
5       env = {};
6       B = {command, motion, msg_distance, led_rgb};
7           command = {{e, e, d_all};
8               < e->e, e->e, d_all<->e >,
9
10              # if distance is small then randomly chose
11              # a new movement direction
12              # 50% straight (green led)
13              < e->c_G, e->m_S, e<->S_all >,
14              < e->c_G, e->m_S, e<->S_all >,
15              # 25% left (red led)
16              < e->c_R, e->m_L, e<->S_all >,
17              # 25% right (blue led)
18              < e->c_B, e->m_R, e<->S_all >,
19
20              # if distance is big then stop
21              < e->c_W, e->m_O, e<->B_all >,
22
23              # publish the new commands
24              < c_G<->e, m_S<->e, S_all->d_all >,
25              < c_R<->e, m_L<->e, S_all->d_all >,
26              < c_B<->e, m_R<->e, S_all->d_all >,
27
28              < c_W<->e, m_O<->e, B_all->d_all >,
29          };
30
31          msg_distance = {{e, e, e};
32              # process distance request
33              < e->e, e->e, e<->d_all >,
34              < e->e, e->e, S_all<->e >,
35              < e->e, e->e, B_all<->e >};
36  };
```

(a)   (b)

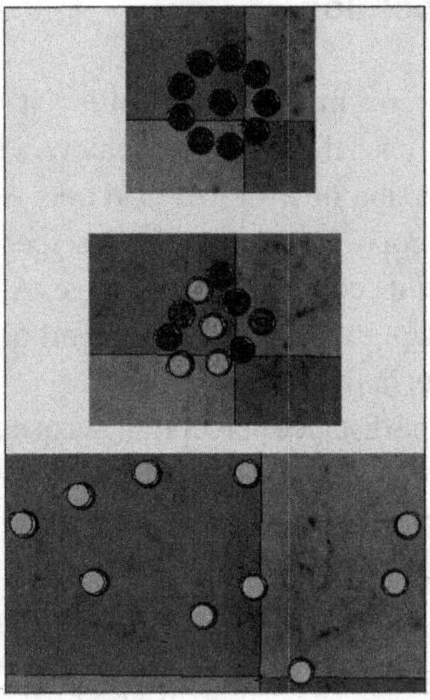

and 25% left or right. By continuously emitting the distance request object (*d_all*), the P colony never halts and can adapt to changes in the distances from neighbors.

The scalability of the *V-REP* simulator is limited (around 20 robots) because of the slow communication line between *Lulu* and *V-REP*. In order to fulfill the scalability requirement more completely, this experiment was also ran on a dedicated *Kilobot* simulator entitled *Kilombo*. This simulator is written in C and together with the C version of *Lulu* as a library, there is no need for a communication channel between the two because now they are joined as a single compiled application. Images from the execution of the dispersion algorithm for 1000 robots using this configuration are presented in Figure 12.

Secure Dispersion

The development process of new emerging technologies generally does not consider the application security as an explicit design objective but rather an afterthought (Higgins, Tomlinson, & Martin, 2009). Because of the defining characteristics of a robot swarm such as mobility, distributed control, autonomy, local communication and emergent behavior, the importance of a security method has been considered from

Figure 12. Screenshots taken during the execution of the P colony dispersion algorithm for 1000 Kilobots using the Kilombo simulator at the initial state (left) and completely dispersed (right)

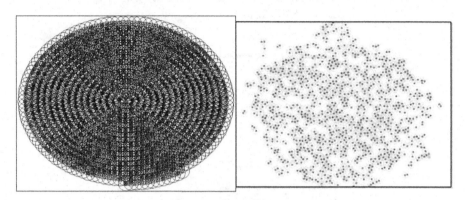

the early stages of algorithm development. All of these characteristics give rise to possible security risks such as: (1) physical capture or blockage, (2) communication disruption or (3) impersonation. Each one of these risks can prevent the swarm from reaching its intended emergent behavior.

Based on the previously discussed dispersion algorithm an improved algorithm is presented that includes an identity detection security method based on a unique number assigned to each robot and was tested on simulated *Kilobot* robots. The basic task of the robots remains the same, to choose a random direction of movement whenever they sense other neighbor robots that are closer than a threshold distance, and so to disperse in the environment.

A section of the dispersion algorithm that includes an ID based method of ignoring intruders is presented in Figure 13 using a P colony representation that can be parsed by the Lulu P colony/P swarm simulator. The entire source code along with demonstration videos are also available at (Florea & Buiu, 2016a). In order to emphasize the control section of the algorithm, the code for the input/output agents has been discarded.

The initial assumption made by the algorithm is that any swarm robot is assigned a unique number that corresponds to a symbolic id object such as *id_2* for robot 2. In the case of *Kilobot* robots, this unique number is stored in the non-volatile memory at calibration time and can be used by applications. In this implementation, all robots execute the same compiled code and determine their symbolic id at startup by subtracting the smallest robot id from the swarm (e.g. 60) from their unique robot id (e.g. 65). In this example, the symbolic id = 65 − 60 = 5. This conversion

Figure 13. Dispersion algorithm with swarm member/non-member identification based on a unique symbolic id used by each swarm member

```
1  pi_disperse = {
2      A = {m_0, m_S, m_L, m_R, c_R, c_G, c_B, c_W, c_0, d_next, B_all, S_*, B_*, id_*};
3      n = 4;
4      env = {id_0, id_1, id_2};
5      B = {command, motion, msg distance, led_rgb};
6          command = {(e, e, e, d_next};
7          # request distance from next robot near me
8          < e->e, e->e, e->e, d_next<->e >,
9
10         # if distance is small then randomly chose a new movement direction
11         # check that the robot id is known (id_* in environment)
12         < e->c_G, e->m_S, e<->id_* / e<->e, e<->S_* >, # 50% straight (green led)
13         < e->c_G, e->m_S, e<->id_* / e<->e, e<->S_* >, # 50% straight (green led)
14         < e->c_R, e->m_L, e<->id_* / e<->e, e<->S_* >, # 25% left (red led)
15         < e->c_B, e->m_R, e<->id_* / e<->e, e<->S_* >, # 25% right (blue led)
16
17         # consume distance big (B_*) objects
18         < e->e, e->e, e->e, e<->B_* >,       # consume step 1
19         < e->e, e->e, e->e, B_*->d_next >,   # consume step 2
20
21         # if distance is big from all robots (B_all in environment) then stop and led white
22         # these are published by msg distance when:
23         # 1) the neighbor robot list is empty
24         # 2) all neighbors are at distance big
25         < e->c_W, e->m_0, e->e, e<->B_all >,
26
27         # consume distance small (S_*) objects from unknown robots
28         # (no id_* taken from environment)
29         # reset motion and led colour commands
30         < c_G->e, m_S->e, e->e, S_*->d_next >,
31         < c_R->e, m_L->e, e->e, S_*->d_next >,
32         < c_B->e, m_R->e, e->e, S_*->d_next >,
33
34         # publish the new commands (given that the robot id is known)
35         # and return the id to the environment for later use
36         < c_G<->e, m_S<->e, id_*<->e, S_*->d_next >,
37         < c_R<->e, m_L<->e, id_*<->e, S_*->d_next >,
38         < c_B<->e, m_R<->e, id_*<->e, S_*->d_next >,
39
40         # publish stop command because there are no close neighbours
41         < c_W<->e, m_0<->e, e->e, B_all->d_next >,
42     };
43 }
```

is necessary because it allows the P colony to be aware of what robot it is being executed on and consequently expand and execute its programs.

For the purpose of increased clarity and error avoidance, a simple wildcard expansion function was implemented within the Lulu P colony simulator and allows the expansion of '*' wildcards with each value from a supplied suffix list. This type of expansion was used in defining the alphabet of the P colony from line 2 in Figure 13, where, for example, the S_* object would be replaced by {S_0, S_1, S_2, S_3, S_4} if this P colony would be executed with a pre-specified swarm size of five robots. The same type of expansion is done for entire programs in an attempt to avoid repeated definitions.

The *message_distance* agent is used to model the infrared distance estimation function, as was the case with the standard dispersion algorithm. Because of the fact that multiple messages can be received in a short amount of time, all messages are buffered internally by the controller application and, upon request (*d_next* object), the agent can respond with one of three objects (S_id, B_id, B_all). These values correspond to the comparison of the current distance to the robot that has the unique symbolic number *id* with a threshold value and have the same meaning of *distance Small* and *distance Big* respectively. If there is no neighbor in the infrared communication range of the robot or all robots are at *distance Big* then the B_all object is emitted in the environment.

The actual security measure is implemented in the code section between lines 12 to 15 and is based on a checking rule (e<->id_* / e<->e) that verifies that the associated *id* object that corresponds to the received *S_id* object exists in the P colony environment. The P colony execution property of randomly choosing an executable program from a set of programs that are executable at one time is used in the same manner as was done for the dispersion algorithm. For example, if the initially declared swarm size is 4 with robots 0, 1, 2, 3 and the environment has the contents listed on line 4 of Figure 13, then if robot 0 encounters robot 3, it will not move away from it because *id_3* is not present in the P colony environment. After checking the id object, it is returned in the environment for later use. This security method is implemented using only P colony programs and does not rely on background security services such as cryptographic functions so as not to move away from the central concept of symbolic control of a robot using membrane computing models.

An example simulation (that used *Kilombo* as base platform) of the dispersion of 9 robots and one intruder is illustrated in Figure 14 with and without the ID security method. In the example on the left side, the intruder (colored using a dark color) is moved close to one of the robots, the "victim" marked using a diamond shape. The "victim" robot sets a new color and starts to move away from the intruder.

Secure Dispersion Pitfalls

The majority, if not all, security methods make several limiting assumptions regarding the context of the algorithm. In what regards the previously discussed global ID

Figure 14. Simulating the dispersion of 9 robots and the effects of the intruder robot (dark body) on the final positions of the robots when running without (a) and with (b) the ID security feature

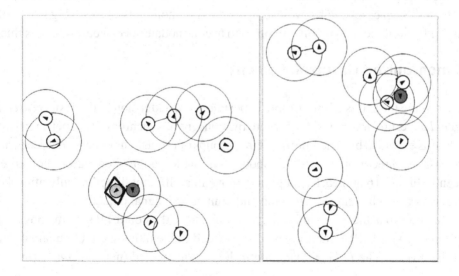

based method of securing the dispersion algorithm, one such limiting assumption is that robots know and are capable of storing all of the IDs of the swarm members. This alone is a limitation in terms of scalability, especially for real *Kilobot* robots that have a RAM memory of only 2Kb.

Another fact that could be considered, not implicitly as a security issue but more as a performance one is the necessity of the algorithm to iterate through all of the neighbor robots in order to determine if one of them is closer than the threshold distance and act upon it. A possible solution could be the use of a request object *d_min* that has a similar meaning to *d_all* in that the former compares the minimum distance of all of the current neighbor robots with the threshold distance and responds using one of the *B_id* or *S_id* depending on the result of the comparison. If there are no robots in the communication range, the internal agent processing function returns the *B_all* object. In essence, this version of the algorithm does an initial sorting of the neighbor distances array in background and compares the threshold distance only with the distance to the closest robot.

A problem that was noticed with this optimized security method was that an intruder robot could block a swarm robot if the intruder was the closest robot to the victim, as seen in Figure 15, where the highlighted robot is the victim and the one colored in magenta is the intruder. The victim cannot start the dispersion from a nearby robot (to the right) because the closest neighbor it has is the intruder and is the only distance measurement returned by the controller backend to the P colony simulator for processing. Because of the security check, the distance measurement for the intruder is ignored and the victim is blocked in a continuous check.

The purpose of this experiment was to emphasize on the importance of an even balance between the logic that is implemented numerically in the backend component of the controller and the symbolic one that is implemented in the frontend, using P colonies. Such cases can be avoided by leveraging more of the processing to one of the ends (numeric or symbolic) in order to accommodate special cases such as this.

Synchronized Group Movement

Continuing the series of multi-robot experiments, an extension of the *P swarm send signal* experiment is introduced where the concept of synchronization between two robots is extended by synchronizing three groups of robots. The experiment scenario consists of three groups of three robots each that are aligned one after another as seen in Figure 16 and each group must move in a different direction only after the previous group has finished moving and emitted the signal object.

In order to achieve group level synchronization, the control P swarm consisted of three types of XP colonies that corresponded to the moving direction and order of the groups. The *Lulu_Kilobot* controller was instructed (using a *config* file) to

Figure 15. A pitfall of the optimized security algorithm where the "victim", the highlighted robot, cannot move away from its closest swarm neighbor on the right

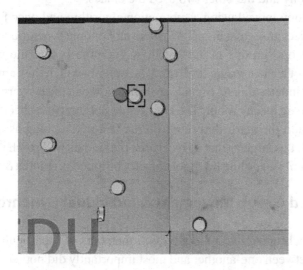

Figure 16. Group synchronization where each group will start moving on a different direction (a), in a top-down order, after the previous group finished moving. Group 1 starts moving to the left when group 0 stopped moving forward (b) and all groups reach the final desired configuration (c)

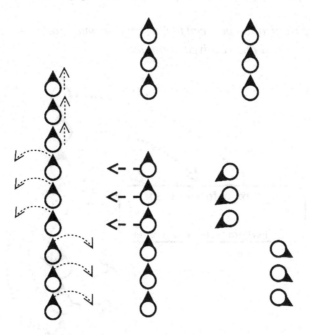

clone each type of XP colony twice where one robot in the group would use the original XP colony and the other two used the clones.

Each XP colony moved using a subtraction algorithm and once finished would emit a signal object as seen in Figure 17 where an example synchronization between the first and second groups in presented. A key note is that the robots are not synchronized (not even at group level) and the order in which they emit and receive signal objects is indeterminate. Signals are broadcasted in the P swarm environment and can be received by any robot. The emitters do not impose a specific destination for the signals and the algorithm relies on the fact that as long as the number of emitters is at least as large as that of receivers, the signals will reach the destination robots. More details regarding the algorithm can be found in (Florea & Buiu, 2016b).

Synchronized Group Movement & Individual Synchronization

Group-level synchronization did not require members of a particular group to be synchronized between one-another and most importantly did not require the notion of individual identity. If, however the task requirements were that after the group synchronization stage, there would be a synchronized individual movement, then there would be a direct need for individual identity. This task requirement is analyzed

Figure 17. Synchronization between the first and second groups: the order in which the signals are emitted and received is random

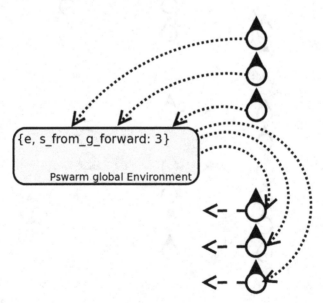

in this section using the same notion of individual identity that was used for the implementation of the security methods.

An extract of the XP colony associated with the group that moves forward is presented in Figure 18. Each XP colony detects the *id* of its associated robot just before runtime, when the expansion of wildcarded objects takes place in the context of the current swarm. For instance, the *my_id_%id* symbolic object would be expanded to *my_id_0* for the first robot from each group. This alternative expansion method is presented graphically in Figure 19 and splits the swarm of robots into three sub-swarms in order to simplify the expansion process especially for cases of non-equal groups of robots and also to simplify the synchronization logic.

Structurally, the XP colonies now have an additional *command* agent, *command_individual* that is triggered into execution after receiving a specific signal object that depends on the symbolic id of the robot. For instance, the first robot from the first group (straight), would wait for a signal (Figure 18 line 9) from the first robot from the first robot from the last group (right) that it had finished moving and individual movement for the first group can start. Similarly, the second and third robots from the first group would wait for a signal (Figure 18, lines 15 and 21) in order to start moving after robot one and two respectively have finished moving. Additional details regarding the structure of the XP colonies and the execution process can be found in (Florea & Buiu, 2016b).

Figure 18. Extract of the XP colony that controls the group of robots moving forward, together as a group as well as individual movement after the group movement stage is over

```
1   xp_group_forward = {
2       A = {1_m, 1_g, g, m_O, m_S, m_L, m_R, c_R, c_G, c_B, s_from_g_forward,
3           s_from_g_right_to_forward_0, my_id_%id, my_id_*, s_internal_%id, s_internal_*, W};
4       n = 3;
5       env = {f, f, f, f, f, f, f, f, f, f, 1_m, my_id_%id, g, g, g, g, g, g, g, g, g, g, g, W };
6       B = {command, command_individual, motion, heartbeat};
7       command = {{e, e, e};
8           # test if I (robot 0 from group_forward) have received a signal from group right
9           < e->individual, e<->s_from_g_right_to_forward_0, e<->my_id_0 >,
10          # if I am robot 0, stop the heartbeat and signal the start of individual movement
11          < s_from_g_right_to_forward_0<->e, individual->individual, my_id_0<->e >,
12
13          # test if I (robot 1 from group_forward) have received a signal
14          # from robot 0 - group_forward
15          < e->individual, e<->s_internal_0, e<->my_id_1 >,
16          # if I am robot 1, stop the heartbeat and signal the start of individual movement
17          < s_internal_0<->e, individual->individual, my_id_1<->e >,
18
19          # test if I (robot 2 from group_forward) have received a signal
20          # from robot 1 - group_forward
21          < e->individual, e<->s_internal_1, e<->my_id_2 >,
22          # if I am robot 2, stop the heartbeat and signal the start of individual movement
23          < s_internal_1<->e, individual->individual, my_id_2<->e >,
24
25          < individual->individual, e->1_g, e->e >,
26          # start the individual movement
27          < individual<->e, 1_g<->e, e->e >,
28
29          # check if we entered individual movement and if so, block this agent
30          < e->e, e<->1_m, e<->individual/e->e >,
31          < 1_m->m_S, e<->>f/e<->e, e->e >,
32          < f->e, m_S<->e, e->e >,
33          < m_S->m_0, e->s_from_g_forward, e->e >,
34          < s_from_g_forward<->e, m_0<->e, e->e >
35      };
36
37      command_individual = {{e, e, e};
38          # the second group movement (move forward, decrease g)
39          < e->e, e<->1_g, e<->individual >,
40          < 1_g->m_S, e<->>g/e<->e, individual<->e >,
41          < g->e, m_S<->e, e->e >,
42          < m_S->m_0, e->s_internal_%id, e->e >,
43          < s_internal_%id<->e, m_0<->e, e->e >
44      };
```

Figure 19. Sub-swarm creation based on the types of XP colonies associated to each group of robots

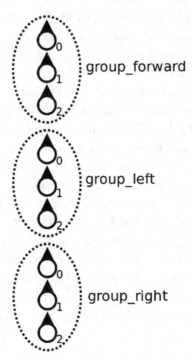

Local Swarm Signal

The locality of direct robot-robot communication is a defining characteristic of a swarm of robots and ensures the use of fully-distributed algorithms by removing any global communication method via a central node. In order to model local communication, the concept of P swarm has been extended into Input/Output P swarm (Figure 20). This concept additionally defines two multisets, *in_global_env* and *out_global_env* alongside the global environment multiset of the P swarm. These two multisets are used in order to achieve an efficient message exchange between an emitter (broadcast) and all receivers that are in his emission range. Among the features of this communication protocol are the simultaneous transmission of multiple symbolic objects (up to 64 objects on the *Kilobot*) and the use of checksums to prevent the multiple receipt of the same message that could disrupt the normal functioning of a XP colony.

Figure 20. The P swarm Input/Output local communication concept applied to Kilobot robots

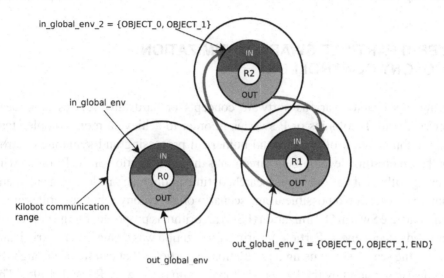

To transmit a message, the emitter places one or more symbolic objects in the *out_global_env* multiset using a special rule type named *out-exteroceptive* (marked with <=O>). After completing the message, the emitter finally adds a finishing symbolic object named *END* that allows the backend of the controller to physically sent the composed message. During reception, the objects are placed in the *in_global_env* multiset and can be read by the agents using *in-exteroceptive* rules (marked with <I=>). The emitter continues to broadcast the message until the *END* object is removed from the *out_global_env* multiset. Because of this fact, receivers always check, using control sums, whether the previously received message and the current one are identical and if so the message is ignored.

Conceptually this local communication protocol is different from the original P swarm global environment in that the latter should be available to all XP colonies at all times. This task can be achieved in a distributed manner but would require a larger communication bandwidth and also time to synchronize the contents of the multiset across all robots. Another advantage of the Input/Output P swarm is that it clearly separates the input and output of an XP colony so as to allow a step by step construction of the output message possibly as a response to the input message.

This local communication protocol allows for the symbolical creation and usage of messages as well as offers a symbolical method of controlling the output device.

HYBRID PARTICLE SWARM OPTIMIZATION: P COLONY CONTROLLER

During this chapter various software concepts or hardware devices have been interfaced with P colonies or P swarms in order to achieve a more complex task using a combination of background numerical processing and symbolic control. Another interesting development direction is the hybridization of the P colony with other algorithms of various types: search, optimization, machine learning and many other. In this section a possible usage scenario of a P colony robot controller along with a Particle Swarm Optimization (PSO) algorithm is proposed, for the purpose of increased effectiveness. Particle Swarm Optimization was designed to heuristically traverse the search space using a population of particles that can move through the search space, guided by the local and global optima (James & Russell, 1995). The current position of a particle is expressed in a coordinate space that has a dimension equal to the number of variables that describe the solution and is changed by adding a velocity that is expressed as a set of real numbers (of equal dimension) (James & Russell, 1995).

The proposed task for the hybrid controller is to decide the role of a robot in a swarm such that it does not conflict with neighboring robots. This problem can be best described as a Graph Coloring (GC) problem. GC is a research topic that is part of the larger domain of Graph Theory (GT) and is concerned with the study of the assignment of colors (labels) to elements of a graph such as vertices, edges and faces such that no two adjacent elements have the same colors, in which case the graph is called *proper* (Axenovich, 2003). The task of determining one or all of the possible combinations of vertex colors has been studied not only in the GT domain but also as a combinatorial problem on which various optimization techniques could be applied in order to reduce the non-polynomial time (NP-hard) needed to check all possible combinations in a recursive manner (Chiarandini & Stützle, 2002).

The GC problem has been used in various domains generally in scheduling applications where tasks are represented as vertices and conflicts are modelled using edges between vertices to prevent tasks from being scheduled simultaneously (Chiarandini & Stützle, 2002). Although the initial definition of the PSO algorithm was based on a global view of the entire swarm of particles, allowing particles to be fully inter-connected, various restricted communication topologies have been introduced, such as the star topology presented in Figure 21, allowing for the development of distributed PSO implementations.

Figure 21. Graph connectivity topologies, fully connected (a), star (b)

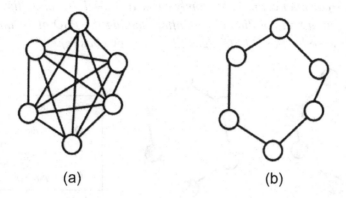

(a)   (b)

Another variant of the PSO, called Binary PSO, focusses on the discretization of the algorithm for use in cases where the target model cannot be represented using real values (position, speed) and instead requires a discrete set of possible values for each variable. The discretization method employed is to use a set of binary values to describe the current position of a particle while the speed is a set of probabilities for these values to become 1 (true) (Kennedy & Eberhart, 1997). This binary variant of PSO is particularly suited for interaction with a P colony because of its discrete nature allowing the use of symbolic objects to describe the current position of a particle or its velocity.

Each robot would start with a random role, marked with a specific color (using the onboard RGB LED and also in the algorithm) and on each iteration of the algorithm, each robot (particle) would choose the best color combination so as not to conflict with neighbor robots, using a topology that is not completely connected in order to respect the swarm robotic guidelines. An example of a typical interaction that could occur between six robots that use a star topology for communications is presented in Figure 22.

CONCLUSION

In this chapter we presented a series of experiments with P colony and P swarm based controllers for single robots and for multiple robots. The experiments show an increasing level of complexity and various control and interaction mechanisms have been accordingly coded in the structure and the rules of the symbolic models based on P colonies and P swarms. A key area of our research in this direction is to use symbolic methods based on the membrane computing principles for approaching various security hazards that a robotic swarm could face. A hybrid system, based on

Figure 22. Swarm robotics example of a task allocation problem where robots (with limited range communications) randomly chose an initial (a) conflicting role (red-red) and obtain a non-conflicting role after running the PSO algorithm (b). The color/role is encoded in the PSO algorithm using two bits (c).

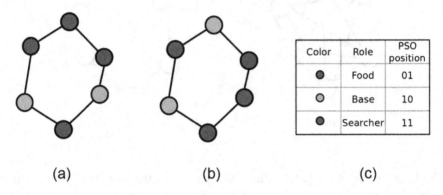

Color	Role	PSO position
●	Food	01
○	Base	10
●	Searcher	11

(a)   (b)   (c)

P colonies and Binary Particle Swarm Optimization is proposed. On the dedicated webpage, the interested reader will find the source code, demonstration videos and instructions for replicating all these experiments on both simulated and real *Kilobot* robots.

REFERENCES

Axenovich, M. (2003). A note on graph coloring extensions and list-colorings. *Electronic Journal of Combinatorics, 10*(1 N), 1–5.

Buiu, C., & Gansari, M. (2014). A new model for interactions between robots in a swarm. In *Electronics, Computers and Artificial Intelligence (ECAI), 2014 6th International Conference on* (pp. 5–10). doi:10.1109/ECAI.2014.7090202

Chiarandini, M., & Stützle, T. (2002). An application of iterated local search to graph coloring problem. In *Proceedings of the Computational Symposium on Graph Coloring and its Generalizations* (pp. 112–125).

Cienciala, L., Ciencialová, L., Csuhaj-Varjú, E., & Vazsil, G. (2010). PCol Automata : Recognizing Strings with P Colonies. *8th Brainstorming Week on Membrane Computing, 1*, 65–76.

Dash, N. P., Dasgupta, R., Chepada, J., & Halder, A. (2011). Event Driven Programming for Embedded Systems - A Finite State Machine Based Approach. In *The Sixth International Conference on Systems (ICONS)* (pp. 19–23).

Florea, A. G., & Buiu, C. (2016a). Demonstration video for the secure dispersion of robots in a swarm using P colonies. http://doi.org/<ALIGNMENT.qj></ALIG NMENT>10.17632/42by9d2f78.1

Florea, A. G., & Buiu, C. (2016b). Synchronized dispersion of robotic swarms using XP colonies. In *Electronics, Computers and Artificial Intelligence (ECAI), 2016 8th Edition International Conference on.*

Heylighen, F. (2016). Stigmergy as a universal coordination mechanism I: Definition and components. *Cognitive Systems Research, 38,* 4–13. doi:10.1016/j. cogsys.2015.12.002

Higgins, F., Tomlinson, A., & Martin, K. M. (2009). Threats to the Swarm : Security Considerations for Swarm Robotics. *International Journal on Advances in Security, 2*(2&3), 288–297. http://doi.org/10.1.1.157.1968

James, K., & Russell, E. (1995). Particle swarm optimization. In *Proceedings of 1995 IEEE International Conference on Neural Networks* (pp. 1942–1948).

Kennedy, J., & Eberhart, R. C. (1997). A discrete binary version of the particle swarm algorithm. *1997 IEEE International Conference on Systems, Man, and Cybernetics. Computational Cybernetics and Simulation, 5,* 4–8. http://doi.org/ doi:10.1109/ICSMC.1997.637339

Langer, M., Cienciala, L., Ciencialová, L., Perdek, M., & Kelemenová, A. (2013). *An Application of the PCol Automata in Robot Control. In 11th Brainstorming Week on Membrane Computing* (pp. 153–164).

Şahin, E. (2005). Swarm Robotics: From Sources of Inspiration to Domains of Application. In *Proceedings of the 2004 International Conference on Swarm Robotics* (pp. 10–20). Springer-Verlag. http://doi.org/ doi:10.1007/978-3-540-30552-1_2

Chapter 5

A Generic Guide for Using Membrane Computing in Robot Control

ABSTRACT

As various theoretical and practical details of using membrane computing models have been presented throughout the book, certain details might be hard to find at a later time. For this reason, this chapter provides the reader with a set of checkmark topics that a developer should address in order to implement a robot controller using a membrane computing model. The topics discussed address areas such as: (1) robot complexity, (2) number of robots, (3) task complexity, (4) simulation versus real world execution, (5) sequential versus parallel implementations. This chapter concludes with an overview of future research directions. These directions offer possible solutions for several important concerns: the development of complex generic algorithms that use a high level of abstraction, the design of swarm algorithms using a top-down (swarm-level) approach and ensuring the predictability of a controller by using concepts such as those used in real-time operating systems.

INTRODUCTION

The use of membrane computing in robot control has gained momentum since the introduction of membrane controllers in (Buiu, Vasile, & Arsene, 2012). New models of P systems amenable to robot control, such as enzymatic P systems, have been proposed. New applications have been approached, such as the control of collective robot systems, including swarms of robots. Furthermore, these approaches have

DOI: 10.4018/978-1-5225-2280-5.ch005

been compared with conventional ones, such as finite-state machines, and some advantages and disadvantages of using membrane computing models in robot control, have been provided.

The aim of this chapter is to offer the curious robotics engineer or intrepid researcher interested in new ways to control robots a guide for selecting the proper membrane computing models and simulators, and what problems could they face in implementing these control models for real-life robots and applications.

This chapter will take the form of a set of guidelines organized along a number of important topics such as:

- Abstraction level;
- Robot complexity;
- The level of interaction between robots;
- Task complexity;
- The type of execution environment, simulated or real-world;
- The level of parallelism.

The majority of the issues are presented in a comparative manner and represent *separate concerns* (a software engineering principle introduced in (Ghezzi, Jazayeri, & Mandrioli, 2002)). These *separate concerns* are important in that a developer should take each one into account when intending to use membrane computing models for robot control.

The following discussion is generic in the sense that it is not centered around a particular scenario but rather tries to accommodate a membrane computing robot controller to different structural and functional requirements.

ABSTRACTION LEVEL

The first fact that must be considered when designing a robot controller is the target level of control or in other words, where will the controller be placed in the control application. This control level is proportional to the abstraction level in the sense that a high level decision application (path planner, role assignment, …) uses only abstract elements as opposed to an embedded device controller (driver) that uses mainly numerical values in order to interface with the hardware. Several example programs are presented graphically in Figure 1.

This type of stacked architecture is often used for operating systems where hardware devices (CPU, memory, …) reside at the bottom of the stack, followed by

Figure 1. Example programs and their corresponding abstraction level

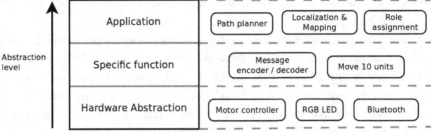

the operating system kernel. A set of system libraries follow immediately above and the interaction with the kernel is done using an *Application Binary Interface* (ABI) while the calls to the system library methods are made by user applications, found at the top level, through an *Application Programming Interface* (API) (Stallings, 2011). For both robot controllers and operating systems, the information flow is bidirectional and generally does not step over any level. Exceptions such as device drivers can be encountered both as part of operating systems but also robot controllers.

Depending on the requirements of the target application, the envisioned controller can be integrated into existing robotics pseudo-operating systems. These software packages are not real operating systems, in the sense of handling memory, CPU, process scheduling, but are actually *Robotic Software Frameworks* (RSF) that provide hardware abstraction for devices commonly found on robots and in some cases also include specialized algorithms. A detailed review and comparison of the most influential RSFs that are also suitable for multi-agent applications is presented in (Iñigo-Blasco, Diaz-del-Rio, Romero-Ternero, Cagigas-Muñiz, & Vicente-Diaz, 2012).

Another important factor in this context, especially for controllers placed at lower abstraction levels, are the performance requirements (most often execution time) that must be met by the membrane controller. If for example, a membrane controller where to be used for the modelling of a high bandwidth communication device, the execution time elapsed between receiving a request and the actual message dispatch should be tested to be small enough so as not to reduce the expected data rate.

FROM SIMPLE TO COMPLEX ROBOTS

The complexity of a robot can usually be divided from several points of view in: mechanical, electronic, and logical complexity (Siegwart & Nourbakhsh, 2004). Each type of complexity is reflected in the controller application at various abstraction levels and is usually handled by increasing the computational power and memory capabilities of the robot. Robots that are simple from all three perspectives, such as the *Kilobot* robot, can be easily controlled with a single microcontroller that runs at a frequency of 8 Mhz and has only 2 Kb of RAM memory (Rubenstein, Ahler, Hoff, Cabrera, & Nagpal, 2014). However, the most constraining factor in this case is the small size of program memory that is used to store the user application, of only 32 Kb, in that it severely limits the length and complexity of programs. This implies that part of the development process of the membrane computing based control application will be reserved to algorithm optimization. The main advantage of such a robotic platform is the simplicity of the control model, where individual devices have a small number of possible states and there are few input sources. For this type of devices, symbolic control of the entire robot is feasible as was shown in Chapter 4.

On the other end of the scale, complex robotic platforms such as humanoid robots present a large set of sensors, have joints and actuators that have multiple degrees of freedom and require higher level control logic. To meet these requirements, for some robots such as the *Nao* humanoid robot, the tasks are split between a microcontroller and a general CPU based on the complexity and abstraction levels. These robots usually have a well-defined control hierarchy and each hardware device is exposed indirectly to the programmer through various library functions. This simplifies the controller design by placing it on top of existing libraries, at a high abstraction level where both symbolic and numeric membrane computing models can be used. Numerical models (such as the enzymatic numerical P systems) are more suitable for some of the tasks that these platforms must perform because they allow for direct numeric control of the position of a joint, on all three axes simultaneously, removing the need for any symbolic to numeric conversions.

The advantages of both numerical and symbolic computing models can be harnessed, as seen in Figure 2, by means of using symbolic models for high level tasks that require reasoning and/or organization and are already expressed using symbols while the numerical membrane models are gradually introduced as the abstraction level decreases (Buiu, 2009).

FROM SINGLE TO MULTIPLE ROBOTS

Depending on the application domain, one or more robots may be required to fulfill a given task. Single robots have to rely on more complex localization, mapping and decisional algorithms in order to follow the requirements without human intervention. In this case, the development of one or more controllers for specialized tasks and devices that belong only to the robot executing the controller are sufficient.

Figure 2. Three tier control architecture and corresponding membrane computing models, as proposed in (Buiu, 2009)

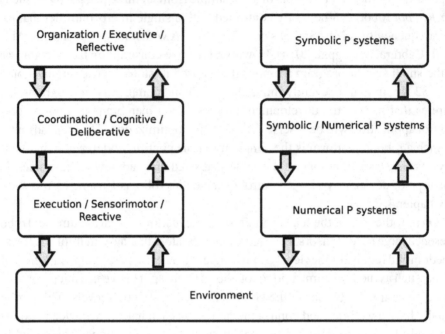

If the intended application of the membrane controller includes robot-to-robot interaction, either direct through a communication protocol or indirect by perceiving the distances from neighbors or the changes they induce in the environment (stigmergy), the notion of neighbors must be included in the design of the controller. Examples of both types of interactions, implemented using P colonies, were presented in Chapter 4. In the case of the *Kilobot* robot, only the distance toward a neighbor can be sensed and so the P colony controller only included a symbolic method of identification of neighbors and of discretizing the distances. A communication protocol was also modelled in later experiments in order to allow the P colony to control the (local) information flow between two or more robots.

As the number of robots increases, so do the scalability requirements of the model that must still be executable even for large numbers of robots. An example in this sense is that of the secure dispersion algorithm that was presented in Chapter 4, *Secure Dispersion*, where the P colony had to store the unique IDs of each robot that was part of the swarm in order to discern them from intruder robots. This also implied that each program that dealt with ID checking had to be written N times if there were N robots in the swarm and caused a depletion of program memory if the example was ran on more than three real *Kilobots*. The most scalable solution (presented graphically in Figure 3 is to not consider global IDs at all and instead

Figure 3. The robot-to-robot interaction model used by the P colony controller that uses global IDs (a) and local IDs (b)

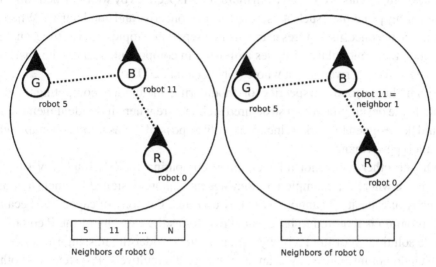

rely on a fixed size list of neighbors in order to locally identify a robot and be able to model its evolution.

In the first example, robot 0 knows and stores the IDs of all robots that are part of the swarm while in the second case, it stores only the index that the robot currently has on the neighbor list. As long as the robots are in communication range, the slot will be reserved but afterwards, it is emptied and all recent knowledge about the neighbor is lost. This solves the scalability issue but raises the difficulty of the symbolic security method that must now infer the identity of the robot from other sources of information (such as receiving a specific message or using a specific synchronization mechanism).

For more complex robots, the same situation occurs only that the communication bandwidth is larger and the robots have more advanced means of perceiving and identifying their neighbors, such as vision (even with depth), sounds and higher processing power that allows the robot to run inverse kinematics algorithms in order to determine an expected position of neighbors.

FROM SIMPLE TO COMPLEX TASKS

The complexity of the task that must be executed by the robot is, from the very beginning, an important design consideration. Difficult tasks generally require, from a software engineering perspective, the development of specialized application components that can be used and reused as needed. This separation of tasks into simpler components is covered from multiple perspectives by well-defined software engineering principles such as separation of concerns and modularity. Whereas separation of concerns requires using distinct separation topics such as time, quality, size and others, modularity implies a division in components that can be analyzed separately and afterwards as a whole (Ghezzi et al., 2002).

From this second perspective of modularity, membrane computing models promote the development of a structured architecture where individual membranes should be designed to work as independently as possible so as to achieve maximum (parallel) performance.

On the other hand, not all membrane computing models implicitly support code reuse where for example P colony agents that are designed for inclusion in a P colony of capacity 2 cannot be directly used in a P colony of capacity 4 because each program is required to have as many rules as the capacity of the P colony. A simple solution is for the P colony simulator to consider the missing rules as *e ->* *e* which do not introduce any change in the agent's multisets. P systems and other related models such as numerical P systems do not present this issue, the only requirement being that the variables in the distribution protocol of the programs are present in the outer membrane.

This concept of modularity and reusability has been extensively used throughout the experiments presented in Chapter 4 where several specialized agents were discussed such as: *motion, led_rgb, timer* and *msg_distance*. These agents were responsible only for interfacing the P colony with the robot hardware. Extended versions of these agents will be developed in order to provide more advanced functions that are repeatedly required in more complex applications and would otherwise increase the complexity of the command agent. For example, an agent that integrated the *motion* agent and the *timer* agent would allow a more complex application to send a request object for moving the robot for 5 seconds in a given direction and then stop.

FROM SIMULATION TO REAL-WORLD

The execution context of robotics applications is important in designing and predicting the expected behavior of the robot. This simulation context can be either a real or simulated robot or group of robots. Both execution environments provide distinct advantages and also disadvantages. The *Kilobot* robot is used as example for the following comparison between simulated and real robots.

According to (Jansson et al., 2015), a swarm of *Kilobot* robots simulated in *Kilombo* provides the benefit of an accelerated algorithm development process by providing a means of simulating 1000 robots at a speed of up to 100 times faster than the real speed. This benefit stems from the fact that the *Kilobot* uses vibration for moving which reduces the speed and precision of movement. This particular simulator also allows the use of the same C source code for both the simulated and real robots in order to reduce platform transition time and also to allow for more accurate comparisons between the evolution of an algorithm on both circumstances. The presence of sensor and effector error was also considered for *Kilombo* as it introduces a Gaussian noise with a configurable probability of receiving a message from a neighbor and also the precision of the estimation.

The real robots have a set of characteristics that were not included in the simulator. The first and most important characteristic is memory space, both RAM and program memory that is limited on the *Kilobot*, as mentioned earlier to 2 Kb and 32 Kb respectively. In the simulator however, this restriction is ignored and the robots are simulated using a portion of the Virtual Memory that is allocated to the process which can be even unlimited, depending on the operating system configuration. This has given rise to situations where the application executes correctly in the simulator but overwrites memory regions on the real robot leading to a possibly undefined behavior. Another important fact that is not modelled accurately by the simulator is the imprecise movement of the robots that is caused by a high sensibility of the vibrating legs to any imperfections on the experiment table.

FROM SEQUENTIAL TO PARALLEL IMPLEMENTATIONS

Execution performance is a constant requirement of most if not all computing applications and this is especially true for robot controllers that must reach certain execution deadlines. Tasks can be grouped into specialized agents, as was discussed throughout this book, and these agents can be executed in two ways: sequential and parallel. The easiest to implement is the sequential mode and if the performance achieved is acceptable, there is no need to pursue a more complicated alternative. Also, depending on the underlying hardware, this could also be the only viable alternative as is the case with lower-cost microcontrollers such as that used for the *Kilobot* robot.

Parallel implementations are achievable using different techniques such as multi-thread and multi-process. This list only considers parallelization techniques that could be employed locally on the robot hardware (provided that it has a multi-task operating system on-board), without resorting to a centralized processing station that could be comprised of a cluster of computers or would feature high performance GPUs for use with graphic card programming languages such as CUDA or OpenCL. A discussion regarding the topic of parallel execution of a P colony has been presented in Chapter 3, *Lulu: An Open-Source Simulator for P Colonies and P Swarms*.

Among the main challenges of such a parallel implementation are the need for exclusive access to global data structures and also the possibility of a deadlock between agents. An example of the former is presented graphically in Figure 4 where two P colony agents are executed in parallel. In this situation, a P colony consisting of only two agents is executed in parallel and the duration of each simulation time step is presented. Write operations on a data structure that can be accessed by all P colony agents, such as the P colony environment, require that the first agent that requests access to this multiset will block access to any other agent (thread or process) until it finishes the write operation. *Agent 2* will subsequently wait and in the parallel execution becomes a sequential execution for a short period of time. This is one example of a typical situation that can result in a decrease of the expected performance of a parallel algorithm and is one of the reasons why careful attention to the execution details of the algorithm is required in order to increase the performance of a (membrane computing) algorithm using parallelization.

CONCLUSION

The development process of a control algorithm in general and especially one that is based on membrane computing usually requires a solution to the various design considerations that were discussed during this chapter. From these design questions, the intended level of abstraction is the most important because it shapes the structure of the algorithm.

All other topics are general checkpoints that often require attention during the development process of robot controllers where either the available hardware/ software, the intended interaction level or the execution platform of the robot can

Figure 4. Parallel execution of two P colony agents that can occasionally require access to the P colony environment (external multiset): Dashed lines separate execution steps

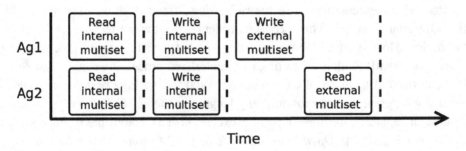

pose several restrictions apart from the complexity level of the tasks or the execution performance requirements.

FUTURE RESEARCH DIRECTIONS

This work has focused on using membrane computing models to build robot controllers that can be either symbolic, numerical or a hierarchical combination of both. As the complexity of robots is growing, there is an inherent need to adapt a control application to various robot configurations. One proposed solution to this problem is to place the controller within a robot control framework such as the *Robot Operating System*. Even though this approach would increase the abstraction level, the advantage would be the increased portability because once the controller is functional on a given robot configuration, only small parameter modifications would be required to adapt the controller to a different configuration.

Another important research direction regards the perspective used to construct the control algorithm for a swarm of robots. Throughout this book, all example controllers were built from the "first-person" perspective of a single robot that could encounter other robots and react according to local sensor perception and message exchanges. This is generally considered as the bottom-up method of constructing a control model because it involves careful planning of the interaction between robots. The opposite top-down perspective allows the user to program the changes that will occur in the space where the robots execute their mission. One example of a programmable space is the *amorphous medium* proposed in (Bachrach, Beal, & McLurkin, 2010) that can be programmed using the *Proto* language and afterwards compiled into individual robot programs. A middleware between the two perspectives is presented in (Pinciroli & Beltrame, 2016) in the form of a swarm-oriented programming language entitled *Buzz*. This language allows robots to be tagged dynamically and the individual robot behavior depends on the current tags. The set of swarm specific constructs is larger but the important fact is that this type of papers set a different trajectory of development, one that could be addressed also using membrane computing models instead of defining a new compiled language.

Robotic systems also have safety requirements, among others, and this justifies the need for a prioritized operating system that can ensure a fast response. This type of operating systems are known as real-time operating systems as they can ensure that a certain task will respect a predefined deadline. In this context, further research will be done in the direction of controller performance evaluation, especially by means of comparison with traditional types of controllers, such as *Finite State Machines*.

REFERENCES

Bachrach, J., Beal, J., & McLurkin, J. (2010). Composable continuous-space programs for robotic swarms. *Neural Computing & Applications*, *19*(6), 825–847. doi:10.1007/s00521-010-0382-8

Buiu, C. (2009). Towards Integrated Biologically Inspired Cognitive Architectures. In *Electronics, Computers and Artificial Intelligence (ECAI), 2009 3rd Edition International Conference on* (pp. 2–8).

Buiu, C., Vasile, C., & Arsene, O. (2012). Development of membrane controllers for mobile robots. *Information Sciences*, *187*, 33–51. doi:10.1016/j.ins.2011.10.007

Ghezzi, C., Jazayeri, M., & Mandrioli, D. (2002). *Fundamentals of software engineering*. Prentice Hall PTR.

Iñigo-Blasco, P., Diaz-del-Rio, F., Romero-Ternero, M. C., Cagigas-Muñiz, D., & Vicente-Diaz, S. (2012). Robotics software frameworks for multi-agent robotic systems development. *Robotics and Autonomous Systems*, *60*(6), 803–821. doi:10.1016/j.robot.2012.02.004

Jansson, F., Hartley, M., Hinsch, M., Slavkov, I., Carranza, N., & Olsson, T. S. G. ... Grieneisen, V. A. (2015). Kilombo: a Kilobot simulator to enable effective research in swarm robotics. *arXiv Preprint arXiv:1511.04285*.

Pinciroli, C., & Beltrame, G. (2016). Swarm-Oriented Programming of Distributed Robot Networks. *Computer*, *49*(12), 32–41. doi:10.1109/MC.2016.376

Rubenstein, M., Ahler, C., Hoff, N., Cabrera, A., & Nagpal, R. (2014). Kilobot: A low cost robot with scalable operations designed for collective behaviors. *Robotics and Autonomous Systems*, *62*(7), 966–975. doi:10.1016/j.robot.2013.08.006

Siegwart, R., & Nourbakhsh, I. R. (2004). *Introduction to Autonomous Mobile Robots* (Vol. 23). Bradford Book.

Stallings, W. (2011). *Operating Systems: Internals and Design Principles*. Prentice Hall.

Related Readings

To continue IGI Global's long-standing tradition of advancing innovation through emerging research, please find below a compiled list of recommended IGI Global book chapters and journal articles in the areas of biologically inspired computing, natural computing, and robotic swarm technology. These related readings will provide additional information and guidance to further enrich your knowledge and assist you with your own research.

Acharjya, D. P., & Kauser, A. P. (2015). Swarm Intelligence in Solving Bio-Inspired Computing Problems: Reviews, Perspectives, and Challenges. In S. Bhattacharyya & P. Dutta (Eds.), *Handbook of Research on Swarm Intelligence in Engineering* (pp. 74–98). Hershey, PA: IGI Global. doi:10.4018/978-1-4666-8291-7.ch003

Acharya, A., & Sinha, D. (2016). A Web-Based Collaborative Learning System Using Concept Maps: Architecture and Evaluation. In J. Mandal, S. Mukhopadhyay, & T. Pal (Eds.), *Handbook of Research on Natural Computing for Optimization Problems* (pp. 916–936). Hershey, PA: IGI Global. doi:10.4018/978-1-5225-0058-2.ch037

Adouane, L. (2016). Flexible and Hybrid Action Selection Process for the Control of Highly Dynamic Multi-Robot Systems. In Y. Tan (Ed.), *Handbook of Research on Design, Control, and Modeling of Swarm Robotics* (pp. 565–595). Hershey, PA: IGI Global. doi:10.4018/978-1-4666-9572-6.ch020

Alam, S., & De, D. (2016). Balanced Energy Consumption Approach Based on Ant Colony in Wireless Sensor Networks. In J. Mandal, S. Mukhopadhyay, & T. Pal (Eds.), *Handbook of Research on Natural Computing for Optimization Problems* (pp. 267–293). Hershey, PA: IGI Global. doi:10.4018/978-1-5225-0058-2.ch012

Ammari, H. M., Shaout, A., & Mustapha, F. (2017). Sensing Coverage in Three-Dimensional Space: A Survey. In N. Ray & A. Turuk (Eds.), *Handbook of Research on Advanced Wireless Sensor Network Applications, Protocols, and Architectures* (pp. 1–28). Hershey, PA: IGI Global. doi:10.4018/978-1-5225-0486-3.ch001

Anandaram, H. (2017). Role of Bioinformatics in Nanotechnology: An Initiation towards Personalized Medicine. In B. Nayak, A. Nanda, & M. Bhat (Eds.), *Integrating Biologically-Inspired Nanotechnology into Medical Practice* (pp. 293–317). Hershey, PA: IGI Global. doi:10.4018/978-1-5225-0610-2.ch011

Ang, L., Zungeru, A. M., Seng, K. P., & Habibi, D. (2015). Artificial Insect Algorithms for Routing in Wireless Sensor Systems. In Y. Shi (Ed.), *Emerging Research on Swarm Intelligence and Algorithm Optimization* (pp. 191–213). Hershey, PA: IGI Global. doi:10.4018/978-1-4666-6328-2.ch009

Anuradha, J., & Tripathy, B. K. (2015). An Uncertainty-Based Model for Optimized Multi-Label Classification. In S. Bhattacharyya & P. Dutta (Eds.), *Handbook of Research on Swarm Intelligence in Engineering* (pp. 40–73). Hershey, PA: IGI Global. doi:10.4018/978-1-4666-8291-7.ch002

Arun, B., & Kumar, T. V. (2015). Materialized View Selection using Marriage in Honey Bees Optimization. *International Journal of Natural Computing Research*, *5*(3), 1–25. doi:10.4018/IJNCR.2015070101

Arun, B., & Kumar, T. V. (2017). Materialized View Selection using Artificial Bee Colony Optimization. *International Journal of Intelligent Information Technologies*, *13*(1), 26–49. doi:10.4018/IJIIT.2017010102

Asawa, K., & Bhardwaj, A. (2016). Sustainability of Public Key Cryptosystem in Quantum Computing Paradigm. In J. Mandal, S. Mukhopadhyay, & T. Pal (Eds.), *Handbook of Research on Natural Computing for Optimization Problems* (pp. 664–688). Hershey, PA: IGI Global. doi:10.4018/978-1-5225-0058-2.ch027

Bagchi, A. (2016). Introduction to Molecular Computation: Theory and Applications – DNA and Membrane Computing. In J. Mandal, S. Mukhopadhyay, & T. Pal (Eds.), *Handbook of Research on Natural Computing for Optimization Problems* (pp. 719–743). Hershey, PA: IGI Global. doi:10.4018/978-1-5225-0058-2.ch029

Bandopadhaya, S., & Roy, J. S. (2017). Spectral Efficiency in Wireless Networks through MIMO-OFDM System. In N. Ray & A. Turuk (Eds.), *Handbook of Research on Advanced Wireless Sensor Network Applications, Protocols, and Architectures* (pp. 249–277). Hershey, PA: IGI Global. doi:10.4018/978-1-5225-0486-3.ch010

Bandyopadhyay, S. K., & Basu, N. (2016). Optimization of Crime Scene Reconstruction Based on Bloodstain Patterns and Machine Learning Techniques. In J. Mandal, S. Mukhopadhyay, & T. Pal (Eds.), *Handbook of Research on Natural Computing for Optimization Problems* (pp. 960–987). Hershey, PA: IGI Global. doi:10.4018/978-1-5225-0058-2.ch039

Banerjee, A., & Ray, S. (2016). An Optimized In Silico Neuroinformatics Approach: Positive Regulation via DNA Interaction in Cellular Decisions for Arg to Ala Mutation in SOX11. In J. Mandal, S. Mukhopadhyay, & T. Pal (Eds.), *Handbook of Research on Natural Computing for Optimization Problems* (pp. 802–820). Hershey, PA: IGI Global. doi:10.4018/978-1-5225-0058-2.ch032

Bar, N., & Das, S. K. (2016). Applicability of ANN in Adsorptive Removal of Cd(II) from Aqueous Solution. In J. Mandal, S. Mukhopadhyay, & T. Pal (Eds.), *Handbook of Research on Natural Computing for Optimization Problems* (pp. 523–560). Hershey, PA: IGI Global. doi:10.4018/978-1-5225-0058-2.ch022

Basu, S. S. (2015). A Self-Organized Software Deployment Architecture for a Swarm Intelligent MANET. In S. Bhattacharyya & P. Dutta (Eds.), *Handbook of Research on Swarm Intelligence in Engineering* (pp. 348–373). Hershey, PA: IGI Global. doi:10.4018/978-1-4666-8291-7.ch011

Benidris, M. A., Elsaiah, S., & Mitra, J. (2015). Applications of Particle Swarm Optimization in Composite Power System Reliability Evaluation. In S. Bhattacharyya & P. Dutta (Eds.), *Handbook of Research on Swarm Intelligence in Engineering* (pp. 573–610). Hershey, PA: IGI Global. doi:10.4018/978-1-4666-8291-7.ch018

Bhattacharya, I. (2016). An Intelligent Approach for Tracking and Monitoring Objects in a Departmental Store Using PSO. In J. Mandal, S. Mukhopadhyay, & T. Pal (Eds.), *Handbook of Research on Natural Computing for Optimization Problems* (pp. 321–338). Hershey, PA: IGI Global. doi:10.4018/978-1-5225-0058-2.ch014

Bhattacharya, S., & Bagchi, A. (2016). Cellular Automata-Basics: Applications in Problem Solving. In J. Mandal, S. Mukhopadhyay, & T. Pal (Eds.), *Handbook of Research on Natural Computing for Optimization Problems* (pp. 619–636). Hershey, PA: IGI Global. doi:10.4018/978-1-5225-0058-2.ch025

Bimonte, S., Sautot, L., Journaux, L., & Faivre, B. (2017). Multidimensional Model Design using Data Mining: A Rapid Prototyping Methodology. *International Journal of Data Warehousing and Mining, 13*(1), 1–35. doi:10.4018/IJDWM.2017010101

Chakraborti, D., Balas, V. E., & Pal, B. B. (2016). Genetic Algorithm for FGP Model of a Multiobjective Bilevel Programming Problem in Uncertain Environment. In J. Mandal, S. Mukhopadhyay, & T. Pal (Eds.), *Handbook of Research on Natural Computing for Optimization Problems* (pp. 870–888). Hershey, PA: IGI Global. doi:10.4018/978-1-5225-0058-2.ch035

Chakraborty, S., & Dey, L. (2016). Image Representation, Filtering, and Natural Computing in a Multivalued Quantum System. In J. Mandal, S. Mukhopadhyay, & T. Pal (Eds.), *Handbook of Research on Natural Computing for Optimization Problems* (pp. 689–717). Hershey, PA: IGI Global. doi:10.4018/978-1-5225-0058-2.ch028

Chandra, S., & Bhattacharyya, S. (2015). Quantum Inspired Swarm Optimization for Multi-Level Image Segmentation Using BDSONN Architecture. In S. Bhattacharyya & P. Dutta (Eds.), *Handbook of Research on Swarm Intelligence in Engineering* (pp. 286–326). Hershey, PA: IGI Global. doi:10.4018/978-1-4666-8291-7.ch009

Charan, B. S., Mittal, A., & Tiwari, R. (2017). Multi-Robot Navigation in Unknown Environment Using Strawberry Algorithm. *International Journal of Robotics Applications and Technologies, 5*(1), 63–81. doi:10.4018/IJRAT.2017010104

Chaudhuri, S. G., & Mukhopadhyaya, K. (2016). Distributed Algorithms for Swarm Robots. In Y. Tan (Ed.), *Handbook of Research on Design, Control, and Modeling of Swarm Robotics* (pp. 207–232). Hershey, PA: IGI Global. doi:10.4018/978-1-4666-9572-6.ch008

Cheng, S., Shi, Y., & Qin, Q. (2015). Experimental Study on Boundary Constraints Handling in Particle Swarm Optimization from a Population Diversity Perspective. In Y. Shi (Ed.), *Emerging Research on Swarm Intelligence and Algorithm Optimization* (pp. 99–127). Hershey, PA: IGI Global. doi:10.4018/978-1-4666-6328-2.ch005

Cheng, S., Shi, Y., & Qin, Q. (2015). Population Diversity of Particle Swarm Optimizer Solving Single- and Multi-Objective Problems. In Y. Shi (Ed.), *Emerging Research on Swarm Intelligence and Algorithm Optimization* (pp. 71–98). Hershey, PA: IGI Global. doi:10.4018/978-1-4666-6328-2.ch004

Costarides, V., Zygomalas, A., Giokas, K., & Koutsouris, D. (2017). Robotics in Surgical Techniques: Present and Future Trends. In A. Moumtzoglou (Ed.), *Design, Development, and Integration of Reliable Electronic Healthcare Platforms* (pp. 86–100). Hershey, PA: IGI Global. doi:10.4018/978-1-5225-1724-5.ch005

Couceiro, M. S. (2016). An Overview of Swarm Robotics for Search and Rescue Applications. In Y. Tan (Ed.), *Handbook of Research on Design, Control, and Modeling of Swarm Robotics* (pp. 345–382). Hershey, PA: IGI Global. doi:10.4018/978-1-4666-9572-6.ch013

Dahal, S., & Ray, N. K. (2017). Intrusion Detection in MANET for Network Layer. In N. Ray & A. Turuk (Eds.), *Handbook of Research on Advanced Wireless Sensor Network Applications, Protocols, and Architectures* (pp. 326–352). Hershey, PA: IGI Global. doi:10.4018/978-1-5225-0486-3.ch013

Das, A., & Chaudhuri, A. (2015). Derivation and Simulation of an Efficient QoS Scheme in MANET through Optimised Messaging Based on ABCO Using QualNet. In S. Bhattacharyya & P. Dutta (Eds.), *Handbook of Research on Swarm Intelligence in Engineering* (pp. 507–536). Hershey, PA: IGI Global. doi:10.4018/978-1-4666-8291-7.ch016

Das, A., Dasgupta, R., & Bagchi, A. (2016). Overview of Cellular Computing-Basic Principles and Applications. In J. Mandal, S. Mukhopadhyay, & T. Pal (Eds.), *Handbook of Research on Natural Computing for Optimization Problems* (pp. 637–662). Hershey, PA: IGI Global. doi:10.4018/978-1-5225-0058-2.ch026

Das, D., Mukhopadhyaya, S., & Nandi, D. (2016). Techniques in Multi-Robot Area Coverage: A Comparative Survey. In Y. Tan (Ed.), *Handbook of Research on Design, Control, and Modeling of Swarm Robotics* (pp. 741–765). Hershey, PA: IGI Global. doi:10.4018/978-1-4666-9572-6.ch027

Das, N., Basu, S., Kundu, M., & Nasipuri, M. (2015). Ambiguity Reduction through Optimal Set of Region Selection Using GA and BFO for Handwritten Bangla Character Recognition. In S. Bhattacharyya & P. Dutta (Eds.), *Handbook of Research on Swarm Intelligence in Engineering* (pp. 611–639). Hershey, PA: IGI Global. doi:10.4018/978-1-4666-8291-7.ch019

Das Sharma, K. (2016). A Comparison among Multi-Agent Stochastic Optimization Algorithms for State Feedback Regulator Design of a Twin Rotor MIMO System. In J. Mandal, S. Mukhopadhyay, & T. Pal (Eds.), *Handbook of Research on Natural Computing for Optimization Problems* (pp. 409–448). Hershey, PA: IGI Global. doi:10.4018/978-1-5225-0058-2.ch018

Dasgupta, K. (2016). Particle Swarm Optimization (PSO) for Optimization in Video Steganography. In J. Mandal, S. Mukhopadhyay, & T. Pal (Eds.), *Handbook of Research on Natural Computing for Optimization Problems* (pp. 339–362). Hershey, PA: IGI Global. doi:10.4018/978-1-5225-0058-2.ch015

Dasgupta, P. (2015). Coverage Path Planning Using Mobile Robot Team Formations. In Y. Shi (Ed.), *Emerging Research on Swarm Intelligence and Algorithm Optimization* (pp. 214–247). Hershey, PA: IGI Global. doi:10.4018/978-1-4666-6328-2.ch010

Datta, N. S., Dutta, H. S., & Majumder, K. (2016). Application of Fuzzy Logic and Fuzzy Optimization Techniques in Medical Image Processing. In J. Mandal, S. Mukhopadhyay, & T. Pal (Eds.), *Handbook of Research on Natural Computing for Optimization Problems* (pp. 822–846). Hershey, PA: IGI Global. doi:10.4018/978-1-5225-0058-2.ch033

Dey, N., & Ashour, A. (2016). *Classification and Clustering in Biomedical Signal Processing* (pp. 1–463). Hershey, PA: IGI Global. doi:10.4018/978-1-5225-0140-4

Dey, S., Bhattacharyya, S., & Maulik, U. (2015). Quantum Behaved Swarm Intelligent Techniques for Image Analysis: A Detailed Survey. In S. Bhattacharyya & P. Dutta (Eds.), *Handbook of Research on Swarm Intelligence in Engineering* (pp. 1–39). Hershey, PA: IGI Global. doi:10.4018/978-1-4666-8291-7.ch001

Ghorai, C., Debnath, A., & Das, A. (2015). A Uniformly Distributed Mobile Sensor Nodes Deployment Strategy Using Swarm Intelligence. In S. Bhattacharyya & P. Dutta (Eds.), *Handbook of Research on Swarm Intelligence in Engineering* (pp. 537–572). Hershey, PA: IGI Global. doi:10.4018/978-1-4666-8291-7.ch017

Ghosal, S. K., & Mandal, J. K. (2016). Genetic-Algorithm-Based Optimization of Fragile Watermarking in Discrete Hartley Transform Domain. In J. Mandal, S. Mukhopadhyay, & T. Pal (Eds.), *Handbook of Research on Natural Computing for Optimization Problems* (pp. 103–127). Hershey, PA: IGI Global. doi:10.4018/978-1-5225-0058-2.ch005

Guler, S., Fidan, B., & Gazi, V. (2016). *Adaptive Swarm Coordination and Formation Control* (Y. Tan, Ed.). doi:10.4018/978-1-4666-9572-6.ch007

Hai-Jew, S. (2017). Finding Automated (Bot, Sensor) or Semi-Automated (Cyborg) Social Media Accounts Using Network Analysis and NodeXL Basic. In N. Rao (Ed.), *Social Media Listening and Monitoring for Business Applications* (pp. 383–424). Hershey, PA: IGI Global. doi:10.4018/978-1-5225-0846-5.ch014

Halder, T., Karforma, S., & Halder, R. (2016). A Secure Data-Hiding Approach Using Particle Swarm Optimization and Pixel Value Difference. In J. Mandal, S. Mukhopadhyay, & T. Pal (Eds.), *Handbook of Research on Natural Computing for Optimization Problems* (pp. 363–381). Hershey, PA: IGI Global. doi:10.4018/978-1-5225-0058-2.ch016

Ion, A., & Patrascu, M. (2017). Agent Based Modelling of Smart Structures: The Challenges of a New Research Domain. In P. Samui, S. Chakraborty, & D. Kim (Eds.), *Modeling and Simulation Techniques in Structural Engineering* (pp. 38–60). Hershey, PA: IGI Global. doi:10.4018/978-1-5225-0588-4.ch002

Jain, U., Godfrey, W. W., & Tiwari, R. (2017). A Hybridization of Gravitational Search Algorithm and Particle Swarm Optimization for Odor Source Localization. *International Journal of Robotics Applications and Technologies*, *5*(1), 20–33. doi:10.4018/IJRAT.2017010102

Janecek, A., & Tan, Y. (2015). Swarm Intelligence for Dimensionality Reduction: How to Improve the Non-Negative Matrix Factorization with Nature-Inspired Optimization Methods. In Y. Shi (Ed.), *Emerging Research on Swarm Intelligence and Algorithm Optimization* (pp. 285–309). Hershey, PA: IGI Global. doi:10.4018/978-1-4666-6328-2.ch013

Jena, G. C. (2017). Multi-Sensor Data Fusion (MSDF). In N. Ray & A. Turuk (Eds.), *Handbook of Research on Advanced Wireless Sensor Network Applications, Protocols, and Architectures* (pp. 29–61). Hershey, PA: IGI Global. doi:10.4018/978-1-5225-0486-3.ch002

Jia, Y. (2016). Design and Implementation for Controlling Multiple Robotic Systems by a Single Operator under Random Communication Delays. In Y. Tan (Ed.), *Handbook of Research on Design, Control, and Modeling of Swarm Robotics* (pp. 672–685). Hershey, PA: IGI Global. doi:10.4018/978-1-4666-9572-6.ch024

Joordens, M., & Champion, B. (2016). Underwater Swarm Robotics: Challenges and Opportunities. In Y. Tan (Ed.), *Handbook of Research on Design, Control, and Modeling of Swarm Robotics* (pp. 718–740). Hershey, PA: IGI Global. doi:10.4018/978-1-4666-9572-6.ch026

Kasemsap, K. (2017). Robotics: Theory and Applications. In M. Moore (Ed.), *Cybersecurity Breaches and Issues Surrounding Online Threat Protection* (pp. 311–345). Hershey, PA: IGI Global. doi:10.4018/978-1-5225-1941-6.ch013

Khoa, T. H., Vasant, P. M., Singh, B. S., & Dieu, V. N. (2015). Swarm-Based Mean-Variance Mapping Optimization (MVMOS) for Solving Non-Convex Economic Dispatch Problems. In S. Bhattacharyya & P. Dutta (Eds.), *Handbook of Research on Swarm Intelligence in Engineering* (pp. 211–251). Hershey, PA: IGI Global. doi:10.4018/978-1-4666-8291-7.ch007

Khurana, D. K., Kapur, P., & Sachdeva, N. (2017). Utility based Tool to Assess Overall Effectiveness of HRD Instruments. *International Journal of Business Analytics*, 4(2), 20–36. doi:10.4018/IJBAN.2017040102

Kirichek, R., Paramonov, A., Vladyko, A., & Borisov, E. (2016). Implementation of the Communication Network for the Multi-Agent Robotic Systems. *International Journal of Embedded and Real-Time Communication Systems*, 7(1), 48–63. doi:10.4018/IJERTCS.2016010103

Klapuch, B. (2017). Trading Orders Algorithm Development: Expert System Approach. In E. Volna, M. Kotyrba, & M. Janosek (Eds.), *Pattern Recognition and Classification in Time Series Data* (pp. 107–126). Hershey, PA: IGI Global. doi:10.4018/978-1-5225-0565-5.ch005

Klepac, G. (2015). Particle Swarm Optimization Algorithm as a Tool for Profiling from Predictive Data Mining Models. In S. Bhattacharyya & P. Dutta (Eds.), *Handbook of Research on Swarm Intelligence in Engineering* (pp. 406–434). Hershey, PA: IGI Global. doi:10.4018/978-1-4666-8291-7.ch013

Klepac, G. (2016). Customer Profiling in Complex Analytical Environments Using Swarm Intelligence Algorithms. *International Journal of Swarm Intelligence Research*, 7(3), 43–70. doi:10.4018/IJSIR.2016070103

Kuila, P., & Jana, P. K. (2016). Evolutionary Computing Approaches for Clustering and Routing in Wireless Sensor Networks. In J. Mandal, S. Mukhopadhyay, & T. Pal (Eds.), *Handbook of Research on Natural Computing for Optimization Problems* (pp. 246–266). Hershey, PA: IGI Global. doi:10.4018/978-1-5225-0058-2.ch011

Kulkarni, P. D., & Ade, R. (2016). Learning from Unbalanced Stream Data in Non-Stationary Environments Using Logistic Regression Model: A Novel Approach Using Machine Learning for Assessment of Credit Card Frauds. In J. Mandal, S. Mukhopadhyay, & T. Pal (Eds.), *Handbook of Research on Natural Computing for Optimization Problems* (pp. 561–582). Hershey, PA: IGI Global. doi:10.4018/978-1-5225-0058-2.ch023

Kumar, M., Balas, V. E., & Pal, B. B. (2016). Using Fuzzy Goal Programming with Penalty Functions for Solving EEPGD Problem via Genetic Algorithm. In J. Mandal, S. Mukhopadhyay, & T. Pal (Eds.), *Handbook of Research on Natural Computing for Optimization Problems* (pp. 847–869). Hershey, PA: IGI Global. doi:10.4018/978-1-5225-0058-2.ch034

Kumar, N., & Singh, Y. (2017). Routing Protocols in Wireless Sensor Networks. In N. Ray & A. Turuk (Eds.), *Handbook of Research on Advanced Wireless Sensor Network Applications, Protocols, and Architectures* (pp. 86–128). Hershey, PA: IGI Global. doi:10.4018/978-1-5225-0486-3.ch004

Kumar, S., Datta, D., & Singh, S. K. (2015). Swarm Intelligence for Biometric Feature Optimization. In S. Bhattacharyya & P. Dutta (Eds.), *Handbook of Research on Swarm Intelligence in Engineering* (pp. 147–181). Hershey, PA: IGI Global. doi:10.4018/978-1-4666-8291-7.ch005

Laing, T. M., Ng, S., Tomlinson, A., & Martin, K. M. (2016). Security in Swarm Robotics. In Y. Tan (Ed.), *Handbook of Research on Design, Control, and Modeling of Swarm Robotics* (pp. 42–66). Hershey, PA: IGI Global. doi:10.4018/978-1-4666-9572-6.ch002

Lapkova, D., Oplatkova, Z. K., Pluhacek, M., Senkerik, R., & Adamek, M. (2017). Analysis and Classification Tools for Automatic Process of Punches and Kicks Recognition. In E. Volna, M. Kotyrba, & M. Janosek (Eds.), *Pattern Recognition and Classification in Time Series Data* (pp. 127–151). Hershey, PA: IGI Global. doi:10.4018/978-1-5225-0565-5.ch006

Lazar, I., Krayem, S., & Hrušecká, D. (2017). Distribution Signals between the Transmitter and Antenna – Event B Model: Distribution TV Signal. In E. Volna, M. Kotyrba, & M. Janosek (Eds.), *Pattern Recognition and Classification in Time Series Data* (pp. 179–217). Hershey, PA: IGI Global. doi:10.4018/978-1-5225-0565-5.ch008

Leong, W. F., Wu, Y., & Yen, G. G. (2015). A Particle Swarm Optimizer for Constrained Multiobjective Optimization. In Y. Shi (Ed.), *Emerging Research on Swarm Intelligence and Algorithm Optimization* (pp. 128–159). Hershey, PA: IGI Global. doi:10.4018/978-1-4666-6328-2.ch006

Li, W., & Tian, Y. (2016). Chemical Plume Tracing and Mapping via Swarm Robots. In Y. Tan (Ed.), *Handbook of Research on Design, Control, and Modeling of Swarm Robotics* (pp. 421–455). Hershey, PA: IGI Global. doi:10.4018/978-1-4666-9572-6.ch016

MacLennan, B. J. (2014). Coordinating Massive Robot Swarms. *International Journal of Robotics Applications and Technologies*, 2(2), 1–19. doi:10.4018/IJRAT.2014070101

Mahmud, N. (2015). Improving Dependability of Robotics Systems, Experience from Application of Fault Tree Synthesis to Analysis of Transport Systems. *International Journal of Robotics Applications and Technologies*, 3(2), 38–62. doi:10.4018/IJRAT.2015070103

Maiti, P., Addya, S. K., Sahoo, B., & Turuk, A. K. (2017). Energy Efficient Wireless Body Area Network (WBAN). In N. Ray & A. Turuk (Eds.), *Handbook of Research on Advanced Wireless Sensor Network Applications, Protocols, and Architectures* (pp. 413–432). Hershey, PA: IGI Global. doi:10.4018/978-1-5225-0486-3.ch017

Majumder, K., De, D., Kar, S., & Singh, R. (2016). Genetic-Algorithm-Based Optimization of Clustering in Mobile Ad Hoc Network. In J. Mandal, S. Mukhopadhyay, & T. Pal (Eds.), *Handbook of Research on Natural Computing for Optimization Problems* (pp. 128–158). Hershey, PA: IGI Global. doi:10.4018/978-1-5225-0058-2.ch006

Malysz, P., & Sirouspour, S. (2014). Mixed Autonomous/Teleoperation Control of Asymmetric Robotic Systems. *International Journal of Robotics Applications and Technologies*, 2(1), 35–60. doi:10.4018/ijrat.2014010103

Manohari, P. K., & Ray, N. K. (2017). A Comprehensive Study of Security in Cloud Computing. In N. Ray & A. Turuk (Eds.), *Handbook of Research on Advanced Wireless Sensor Network Applications, Protocols, and Architectures* (pp. 386–412). Hershey, PA: IGI Global. doi:10.4018/978-1-5225-0486-3.ch016

Martinez-Martin, E., & del Pobil, A. P. (2016). Conflict Resolution in Robotics: An Overview. In P. Novais & D. Carneiro (Eds.), *Interdisciplinary Perspectives on Contemporary Conflict Resolution* (pp. 263–278). Hershey, PA: IGI Global. doi:10.4018/978-1-5225-0245-6.ch015

Meghanathan, N. (2017). Impact of the Structure of the Data Gathering Trees on Node Lifetime and Network Lifetime in Wireless Sensor Networks. In N. Ray & A. Turuk (Eds.), *Handbook of Research on Advanced Wireless Sensor Network Applications, Protocols, and Architectures* (pp. 184–196). Hershey, PA: IGI Global. doi:10.4018/978-1-5225-0486-3.ch007

Mendes, R. P., Calado, M. D., & Mariano, S. J. (2015). Particle Swarm Optimization Method to Design a Linear Tubular Switched Reluctance Generator. In S. Bhattacharyya & P. Dutta (Eds.), *Handbook of Research on Swarm Intelligence in Engineering* (pp. 469–506). Hershey, PA: IGI Global. doi:10.4018/978-1-4666-8291-7.ch015

Meralto, C., Moura, J., & Marinheiro, R. (2017). Wireless Mesh Sensor Networks with Mobile Devices: A Comprehensive Review. In N. Ray & A. Turuk (Eds.), *Handbook of Research on Advanced Wireless Sensor Network Applications, Protocols, and Architectures* (pp. 129–155). Hershey, PA: IGI Global. doi:10.4018/978-1-5225-0486-3.ch005

Mishra, A. K. (2017). Security Threats in Wireless Sensor Networks. In N. Ray & A. Turuk (Eds.), *Handbook of Research on Advanced Wireless Sensor Network Applications, Protocols, and Architectures* (pp. 307–325). Hershey, PA: IGI Global. doi:10.4018/978-1-5225-0486-3.ch012

Mishra, R., & Das, K. N. (2017). A Novel Hybrid Genetic Algorithm for Unconstrained and Constrained Function Optimization. In D. Acharjya & A. Mitra (Eds.), *Bio-Inspired Computing for Information Retrieval Applications* (pp. 230–268). Hershey, PA: IGI Global. doi:10.4018/978-1-5225-2375-8.ch009

Mishra, S., Mishra, B. K., & Tripathy, H. K. (2017). Significance of Biologically Inspired Optimization Techniques in Real-Time Applications. In D. Acharjya & A. Mitra (Eds.), *Bio-Inspired Computing for Information Retrieval Applications* (pp. 150–180). Hershey, PA: IGI Global. doi:10.4018/978-1-5225-2375-8.ch006

Mishra, S., Singh, S. S., Mishra, B. S., & Panigrahi, P. K. (2016). Research on Soft Computing Techniques for Cognitive Radio. *International Journal of Mobile Computing and Multimedia Communications*, 7(2), 53–73. doi:10.4018/IJMCMC.2016040104

Mo, H., Xu, L., & Geng, M. (2015). Image Segmentation Based on Bio-Inspired Optimization Algorithms. In Y. Shi (Ed.), *Emerging Research on Swarm Intelligence and Algorithm Optimization* (pp. 259–284). Hershey, PA: IGI Global. doi:10.4018/978-1-4666-6328-2.ch012

Mohanty, J. P., & Mandal, C. (2017). Connected Dominating Set in Wireless Sensor Network. In N. Ray & A. Turuk (Eds.), *Handbook of Research on Advanced Wireless Sensor Network Applications, Protocols, and Architectures* (pp. 62–85). Hershey, PA: IGI Global. doi:10.4018/978-1-5225-0486-3.ch003

Mohanty, S., & Patra, S. K. (2017). Performance Evaluation of Quality of Service in IEEE 802.15.4-Based Wireless Sensor Networks. In N. Ray & A. Turuk (Eds.), *Handbook of Research on Advanced Wireless Sensor Network Applications, Protocols, and Architectures* (pp. 213–248). Hershey, PA: IGI Global. doi:10.4018/978-1-5225-0486-3.ch009

Mondal, M. A., & Deb, T. (2016). Optimized Energy Aware VM Provisioning in Green Cloud Based on Cuckoo Search with Levy Flight. In J. Mandal, S. Mukhopadhyay, & T. Pal (Eds.), *Handbook of Research on Natural Computing for Optimization Problems* (pp. 449–474). Hershey, PA: IGI Global. doi:10.4018/978-1-5225-0058-2.ch019

Mukherjee, A., Deb, P., & De, D. (2016). Natural Computing in Mobile Network Optimization. In J. Mandal, S. Mukhopadhyay, & T. Pal (Eds.), *Handbook of Research on Natural Computing for Optimization Problems* (pp. 382–408). Hershey, PA: IGI Global. doi:10.4018/978-1-5225-0058-2.ch017

Mukhopadhyay, A. (2016). MRI Brain Image Segmentation Using Interactive Multiobjective Evolutionary Approach. In J. Mandal, S. Mukhopadhyay, & T. Pal (Eds.), *Handbook of Research on Natural Computing for Optimization Problems* (pp. 10–29). Hershey, PA: IGI Global. doi:10.4018/978-1-5225-0058-2.ch002

Mukhopadhyay, S., & Das, S. (2016). A System on Chip Development of Customizable GA Architecture for Real Parameter Optimization Problem. In J. Mandal, S. Mukhopadhyay, & T. Pal (Eds.), *Handbook of Research on Natural Computing for Optimization Problems* (pp. 66–102). Hershey, PA: IGI Global. doi:10.4018/978-1-5225-0058-2.ch004

Mukhopadhyay, S., Mandal, J. K., & Pal, T. (2016). Variable Length PSO-Based Image Clustering for Image Denoising. In J. Mandal, S. Mukhopadhyay, & T. Pal (Eds.), *Handbook of Research on Natural Computing for Optimization Problems* (pp. 294–320). Hershey, PA: IGI Global. doi:10.4018/978-1-5225-0058-2.ch013

Naik, B., Nayak, J., & Behera, H. S. (2016). A Hybrid Model of FLANN and Firefly Algorithm for Classification. In J. Mandal, S. Mukhopadhyay, & T. Pal (Eds.), *Handbook of Research on Natural Computing for Optimization Problems* (pp. 491–522). Hershey, PA: IGI Global. doi:10.4018/978-1-5225-0058-2.ch021

Neogi, A. (2015). Studies of Computational Intelligence Based on the Behaviour of Cockroaches. In S. Bhattacharyya & P. Dutta (Eds.), *Handbook of Research on Swarm Intelligence in Engineering* (pp. 99–146). Hershey, PA: IGI Global. doi:10.4018/978-1-4666-8291-7.ch004

Nunnally, S., Walker, P., Lewis, M., Chakraborty, N., & Sycara, K. (2016). Using Haptic Feedback in Human-Swarm Interaction. In Y. Tan (Ed.), *Handbook of Research on Design, Control, and Modeling of Swarm Robotics* (pp. 619–644). Hershey, PA: IGI Global. doi:10.4018/978-1-4666-9572-6.ch022

Pal, B. B., Roy, S., & Kumar, M. (2016). A Genetic Algorithm to Goal Programming Model for Crop Production with Interval Data Uncertainty. In J. Mandal, S. Mukhopadhyay, & T. Pal (Eds.), *Handbook of Research on Natural Computing for Optimization Problems* (pp. 30–65). Hershey, PA: IGI Global. doi:10.4018/978-1-5225-0058-2.ch003

Pan, I., & Samanta, T. (2015). Advanced Strategy for Droplet Routing in Digital Microfluidic Biochips Using ACO. In S. Bhattacharyya & P. Dutta (Eds.), *Handbook of Research on Swarm Intelligence in Engineering* (pp. 252–284). Hershey, PA: IGI Global. doi:10.4018/978-1-4666-8291-7.ch008

Panda, S. (2017). Security Issues and Challenges in Internet of Things. In N. Ray & A. Turuk (Eds.), *Handbook of Research on Advanced Wireless Sensor Network Applications, Protocols, and Architectures* (pp. 369–385). Hershey, PA: IGI Global. doi:10.4018/978-1-5225-0486-3.ch015

Pereira, L. M., & Saptawijaya, A. (2015). Bridging Two Realms of Machine Ethics. In J. White & R. Searle (Eds.), *Rethinking Machine Ethics in the Age of Ubiquitous Technology* (pp. 197–224). Hershey, PA: IGI Global. doi:10.4018/978-1-4666-8592-5.ch010

Politis, D., Tsaligopoulos, M., & Kyriafinis, G. (2016). From Cochlear Implants and Neurotology to Brain Computer Interfaces: Exploring the World of Neuron Synapses for Hearing Impairments. In J. Mandal, S. Mukhopadhyay, & T. Pal (Eds.), *Handbook of Research on Natural Computing for Optimization Problems* (pp. 988–1015). Hershey, PA: IGI Global. doi:10.4018/978-1-5225-0058-2.ch040

Pour Sadrollah, G., Barca, J. C., Eliasson, J., & Khan, A. I. (2016). A Distributed Framework and Consensus Middle-Ware for Human Swarm Interaction. In Y. Tan (Ed.), *Handbook of Research on Design, Control, and Modeling of Swarm Robotics* (pp. 645–671). Hershey, PA: IGI Global. doi:10.4018/978-1-4666-9572-6.ch023

Pratihar, D. K. (2016). Realizing the Need for Intelligent Optimization Tool. In J. Mandal, S. Mukhopadhyay, & T. Pal (Eds.), *Handbook of Research on Natural Computing for Optimization Problems* (pp. 1–9). Hershey, PA: IGI Global. doi:10.4018/978-1-5225-0058-2.ch001

Prusty, A. R., Sethi, S., & Nayak, A. K. (2017). Energy Aware Optimized Routing Protocols for Wireless Ad Hoc Sensor Network. In N. Ray & A. Turuk (Eds.), *Handbook of Research on Advanced Wireless Sensor Network Applications, Protocols, and Architectures* (pp. 156–183). Hershey, PA: IGI Global. doi:10.4018/978-1-5225-0486-3.ch006

Purushothaman, G. (2017). Bio-Inspired Techniques in Rehabilitation Engineering for Control of Assistive Devices. In D. Acharjya & A. Mitra (Eds.), *Bio-Inspired Computing for Information Retrieval Applications* (pp. 293–315). Hershey, PA: IGI Global. doi:10.4018/978-1-5225-2375-8.ch011

Rahman, I., Vasant, P., Singh, B. S., & Abdullah-Al-Wadud, M. (2015). Swarm Intelligence-Based Optimization for PHEV Charging Stations. In S. Bhattacharyya & P. Dutta (Eds.), *Handbook of Research on Swarm Intelligence in Engineering* (pp. 374–405). Hershey, PA: IGI Global. doi:10.4018/978-1-4666-8291-7.ch012

Rathipriya, R., & Thangavel, K. (2015). Hybrid Swarm Intelligence-Based Biclustering Approach for Recommendation of Web Pages. In Y. Shi (Ed.), *Emerging Research on Swarm Intelligence and Algorithm Optimization* (pp. 161–180). Hershey, PA: IGI Global. doi:10.4018/978-1-4666-6328-2.ch007

Ray, S. (2016). Adaptive Simulated Annealing Algorithm to Solve Bio-Molecular Optimization. In J. Mandal, S. Mukhopadhyay, & T. Pal (Eds.), *Handbook of Research on Natural Computing for Optimization Problems* (pp. 475–489). Hershey, PA: IGI Global. doi:10.4018/978-1-5225-0058-2.ch020

Ray, S. (2016). Evolutionary Computing to Examine Variation in Proteins with Evolution. In J. Mandal, S. Mukhopadhyay, & T. Pal (Eds.), *Handbook of Research on Natural Computing for Optimization Problems* (pp. 185–200). Hershey, PA: IGI Global. doi:10.4018/978-1-5225-0058-2.ch008

Rout, S., Turuk, A. K., & Sahoo, B. (2017). Techniques to Enhance the Lifetime of MANET. In N. Ray & A. Turuk (Eds.), *Handbook of Research on Advanced Wireless Sensor Network Applications, Protocols, and Architectures* (pp. 278–306). Hershey, PA: IGI Global. doi:10.4018/978-1-5225-0486-3.ch011

Roy, P., Dey, D., De, D., & Sinha, S. (2016). DNA Cryptography. In J. Mandal, S. Mukhopadhyay, & T. Pal (Eds.), *Handbook of Research on Natural Computing for Optimization Problems* (pp. 775–801). Hershey, PA: IGI Global. doi:10.4018/978-1-5225-0058-2.ch031

Roy, P. K., Pradhan, M., & Pal, T. (2016). Evolutionary Algorithms for Economic Load Dispatch Having Multiple Types of Cost Functions. In J. Mandal, S. Mukhopadhyay, & T. Pal (Eds.), *Handbook of Research on Natural Computing for Optimization Problems* (pp. 201–226). Hershey, PA: IGI Global. doi:10.4018/978-1-5225-0058-2.ch009

Saha, R., & Bhowmik, M. K. (2016). Active Contour Model for Medical Applications. In J. Mandal, S. Mukhopadhyay, & T. Pal (Eds.), *Handbook of Research on Natural Computing for Optimization Problems* (pp. 937–959). Hershey, PA: IGI Global. doi:10.4018/978-1-5225-0058-2.ch038

Said, G. A., & El-Horbaty, E. M. (2016). Optimizing Solution for Storage Space Allocation Problem in Container Terminal Using Genetic Algorithm. In J. Mandal, S. Mukhopadhyay, & T. Pal (Eds.), *Handbook of Research on Natural Computing for Optimization Problems* (pp. 159–184). Hershey, PA: IGI Global. doi:10.4018/978-1-5225-0058-2.ch007

Samui, P., Dalkiliç, Y. H., Rajadurai, H., & Jagan, J. (2015). Minimax Probability Machine: A New Tool for Modeling Seismic Liquefaction Data. In S. Bhattacharyya & P. Dutta (Eds.), *Handbook of Research on Swarm Intelligence in Engineering* (pp. 182–210). Hershey, PA: IGI Global. doi:10.4018/978-1-4666-8291-7.ch006

Santhi, V., & Tripathy, B. K. (2015). Image Enhancement Techniques Using Particle Swarm Optimization Technique. In S. Bhattacharyya & P. Dutta (Eds.), *Handbook of Research on Swarm Intelligence in Engineering* (pp. 327–347). Hershey, PA: IGI Global. doi:10.4018/978-1-4666-8291-7.ch010

Sarkar, A., & Das, R. (2015). Remote Sensing Image Classification Using Fuzzy-PSO Hybrid Approach. In S. Bhattacharyya & P. Dutta (Eds.), *Handbook of Research on Swarm Intelligence in Engineering* (pp. 435–468). Hershey, PA: IGI Global. doi:10.4018/978-1-4666-8291-7.ch014

Sarkar, A., & Mandal, J. K. (2015). Particle Swarm Optimization-Based Session Key Generation for Wireless Communication (PSOSKG). In S. Bhattacharyya & P. Dutta (Eds.), *Handbook of Research on Swarm Intelligence in Engineering* (pp. 640–677). Hershey, PA: IGI Global. doi:10.4018/978-1-4666-8291-7.ch020

Sarma, A., Sarmah, K., Sarma, K. K., Goswami, S., & Baruah, S. (2016). Soft-Computing-Based Optimization of Low Return Loss Multiband Microstrip Patch Antenna. In J. Mandal, S. Mukhopadhyay, & T. Pal (Eds.), *Handbook of Research on Natural Computing for Optimization Problems* (pp. 583–617). Hershey, PA: IGI Global. doi:10.4018/978-1-5225-0058-2.ch024

Satpathy, R. (2017). Bioinspired Algorithms in Solving Three-Dimensional Protein Structure Prediction Problems. In D. Acharjya & A. Mitra (Eds.), *Bio-Inspired Computing for Information Retrieval Applications* (pp. 316–337). Hershey, PA: IGI Global. doi:10.4018/978-1-5225-2375-8.ch012

Sengupta, D., & Chaudhuri, A. (2016). Vedic Sutras: A New Paradigm for Optimizing Arithmetic Operations. In J. Mandal, S. Mukhopadhyay, & T. Pal (Eds.), *Handbook of Research on Natural Computing for Optimization Problems* (pp. 890–915). Hershey, PA: IGI Global. doi:10.4018/978-1-5225-0058-2.ch036

Sethi, S., & Sahoo, R. K. (2017). Design of WSN in Real Time Application of Health Monitoring System. In N. Ray & A. Turuk (Eds.), *Handbook of Research on Advanced Wireless Sensor Network Applications, Protocols, and Architectures* (pp. 197–212). Hershey, PA: IGI Global. doi:10.4018/978-1-5225-0486-3.ch008

Shi, Y. (2015). An Optimization Algorithm Based on Brainstorming Process. In Y. Shi (Ed.), *Emerging Research on Swarm Intelligence and Algorithm Optimization* (pp. 1–35). Hershey, PA: IGI Global. doi:10.4018/978-1-4666-6328-2.ch001

Shylo, V. P., & Shylo, O. V. (2015). Path Relinking Scheme for the Max-Cut Problem. In Y. Shi (Ed.), *Emerging Research on Swarm Intelligence and Algorithm Optimization* (pp. 248–258). Hershey, PA: IGI Global. doi:10.4018/978-1-4666-6328-2.ch011

Silva, L. Jr, & Nedjah, N. (2016). Distributed Algorithms for Recruitment and Coordinated Motion in Swarm Robotic Systems. In Y. Tan (Ed.), *Handbook of Research on Design, Control, and Modeling of Swarm Robotics* (pp. 596–617). Hershey, PA: IGI Global. doi:10.4018/978-1-4666-9572-6.ch021

Singh, J. P., & Dutta, P. (2016). Source Location Privacy Using Ant Colony Optimization in Wireless Sensor Networks. In J. Mandal, S. Mukhopadhyay, & T. Pal (Eds.), *Handbook of Research on Natural Computing for Optimization Problems* (pp. 227–245). Hershey, PA: IGI Global. doi:10.4018/978-1-5225-0058-2.ch010

Singh, V. P., Srivastava, S., & Srivastava, R. (2017). An Efficient Image Retrieval Based on Fusion of Fast Features and Query Image Classification. *International Journal of Rough Sets and Data Analysis*, 4(1), 19–37. doi:10.4018/IJRSDA.2017010102

Sinha, S., Bandyopadhyay, J., & De, D. (2016). DNA Computing Using Carbon Nanotube-DNA Hybrid Nanostructure. In J. Mandal, S. Mukhopadhyay, & T. Pal (Eds.), *Handbook of Research on Natural Computing for Optimization Problems* (pp. 744–774). Hershey, PA: IGI Global. doi:10.4018/978-1-5225-0058-2.ch030

Svejda, J., Zak, R., Senkerik, R., & Jasek, R. (2017). Research on Processing the Brain Activity in BCI System. In E. Volna, M. Kotyrba, & M. Janosek (Eds.), *Pattern Recognition and Classification in Time Series Data* (pp. 152–178). Hershey, PA: IGI Global. doi:10.4018/978-1-5225-0565-5.ch007

Tan, Y. (2016). A Survey on Swarm Robotics. In Y. Tan (Ed.), *Handbook of Research on Design, Control, and Modeling of Swarm Robotics* (pp. 1–41). Hershey, PA: IGI Global. doi:10.4018/978-1-4666-9572-6.ch001

Telnarova, Z. (2017). Modeling and Language Support for the Pattern Management. In E. Volna, M. Kotyrba, & M. Janosek (Eds.), *Pattern Recognition and Classification in Time Series Data* (pp. 86–106). Hershey, PA: IGI Global. doi:10.4018/978-1-5225-0565-5.ch004

Ting, T. O. (2015). Optimization of Drilling Process via Weightless Swarm Algorithm. In Y. Shi (Ed.), *Emerging Research on Swarm Intelligence and Algorithm Optimization* (pp. 181–190). Hershey, PA: IGI Global. doi:10.4018/978-1-4666-6328-2.ch008

Ulbrich, F., Rotter, S. S., & Rojas, R. (2016). Adapting to the Traffic Swarm: Swarm Behaviour for Autonomous Cars. In Y. Tan (Ed.), *Handbook of Research on Design, Control, and Modeling of Swarm Robotics* (pp. 263–285). Hershey, PA: IGI Global. doi:10.4018/978-1-4666-9572-6.ch010

Urrea, C., & Araya, H. (2017). New Redundant Manipulator Robot with Six Degrees of Freedom Controlled with Visual Feedback. In N. Dey, A. Ashour, & P. Patra (Eds.), *Feature Detectors and Motion Detection in Video Processing* (pp. 231–255). Hershey, PA: IGI Global. doi:10.4018/978-1-5225-1025-3.ch011

Ursyn, A. (2015). Duality of Natural and Technological Explanations. In A. Ursyn (Ed.), *Handbook of Research on Maximizing Cognitive Learning through Knowledge Visualization* (pp. 113–199). Hershey, PA: IGI Global. doi:10.4018/978-1-4666-8142-2.ch005

Vallverdú, J., & Talanov, M. (2017). Naturalizing Consciousness Emergence for AI Implementation Purposes: A Guide to Multilayered Management Systems. In J. Vallverdú, M. Mazzara, M. Talanov, S. Distefano, & R. Lowe (Eds.), *Advanced Research on Biologically Inspired Cognitive Architectures* (pp. 24–40). Hershey, PA: IGI Global. doi:10.4018/978-1-5225-1947-8.ch002

Venkatesan, A. S. (2017). Optimized Clustering Techniques with Special Focus to Biomedical Datasets. In B. Singh (Ed.), *Computational Tools and Techniques for Biomedical Signal Processing* (pp. 333–360). Hershey, PA: IGI Global. doi:10.4018/978-1-5225-0660-7.ch015

Volna, E., & Kotyrba, M. (2017). Recognition of Patterns with Fractal Structure in Time Series. In E. Volna, M. Kotyrba, & M. Janosek (Eds.), *Pattern Recognition and Classification in Time Series Data* (pp. 1–31). Hershey, PA: IGI Global. doi:10.4018/978-1-5225-0565-5.ch001

Wang, S., Zhao, Y., Shu, Y., & Shi, W. (2017). Improved Approximation Algorithm for Maximal Information Coefficient. *International Journal of Data Warehousing and Mining, 13*(1), 76–93. doi:10.4018/IJDWM.2017010104

Yan, X., & Sun, D. (2016). Shape Control of Robot Swarms with Multilevel-Based Topology Design. In Y. Tan (Ed.), *Handbook of Research on Design, Control, and Modeling of Swarm Robotics* (pp. 233–261). Hershey, PA: IGI Global. doi:10.4018/978-1-4666-9572-6.ch009

Yang, X. (2015). Analysis of Firefly Algorithms and Automatic Parameter Tuning. In Y. Shi (Ed.), *Emerging Research on Swarm Intelligence and Algorithm Optimization* (pp. 36–49). Hershey, PA: IGI Global. doi:10.4018/978-1-4666-6328-2.ch002

Yao, Y., Marchal, K., & Van de Peer, Y. (2016). Adaptive Self-Organizing Organisms Using A Bio-Inspired Gene Regulatory Network Controller: For the Aggregation of Evolutionary Robots under a Changing Environment. In Y. Tan (Ed.), *Handbook of Research on Design, Control, and Modeling of Swarm Robotics* (pp. 68–82). Hershey, PA: IGI Global. doi:10.4018/978-1-4666-9572-6.ch003

Yasuda, G. (2015). Distributed Coordination Architecture for Cooperative Task Planning and Execution of Intelligent Multi-Robot Systems. In A. Azar & S. Vaidyanathan (Eds.), *Handbook of Research on Advanced Intelligent Control Engineering and Automation* (pp. 407–426). Hershey, PA: IGI Global. doi:10.4018/978-1-4666-7248-2.ch015

Yasuda, G. (2016). Design and Implementation of Distributed Autonomous Coordinators for Cooperative Multi-Robot Systems. *International Journal of System Dynamics Applications*, 5(4), 1–15. doi:10.4018/IJSDA.2016100101

Yin, P., Glover, F., Laguna, M., & Zhu, J. (2015). A Complementary Cyber Swarm Algorithm. In Y. Shi (Ed.), *Emerging Research on Swarm Intelligence and Algorithm Optimization* (pp. 50–70). Hershey, PA: IGI Global. doi:10.4018/978-1-4666-6328-2.ch003

Yuce, B., & Mastrocinque, E. (2016). Supply Chain Network Design Using an Enhanced Hybrid Swarm-Based Optimization Algorithm. In P. Vasant, G. Weber, & V. Dieu (Eds.), *Handbook of Research on Modern Optimization Algorithms and Applications in Engineering and Economics* (pp. 95–112). Hershey, PA: IGI Global. doi:10.4018/978-1-4666-9644-0.ch003

Yuce, B., Mastrocinque, E., Packianather, M. S., Lambiase, A., & Pham, D. T. (2015). The Bees Algorithm and Its Applications. In P. Vasant (Ed.), *Handbook of Research on Artificial Intelligence Techniques and Algorithms* (pp. 122–151). Hershey, PA: IGI Global. doi:10.4018/978-1-4666-7258-1.ch004

Žáček, M. (2017). Introduction to Time Series. In E. Volna, M. Kotyrba, & M. Janosek (Eds.), *Pattern Recognition and Classification in Time Series Data* (pp. 32–52). Hershey, PA: IGI Global. doi:10.4018/978-1-5225-0565-5.ch002

Zedadra, A., Lafifi, Y., & Zedadra, O. (2016). Dynamic Group Formation based on a Natural Phenomenon. *International Journal of Distance Education Technologies*, 14(4), 13–26. doi:10.4018/IJDET.2016100102

About the Authors

Andrei George Florea was born in 1991 and graduated Informatics at the Letters and Sciences Faculty of the Petroleum-Gas University of Ploiesti, Romania. After graduating the Advanced Technologies for Data Processing Master's degree program at the same university, he enrolled in the Ph.D. program entitled *Bioinspired Algorithms for Robotic Swarms Security* at the Department of Automatic Control and Systems Engineering of the Politehnica University of Bucharest, Romania. His primary research interests are in software engineering, swarm intelligence, distributed algorithms and embedded systems.

Cătălin Buiu was born in 1968 and in 1992 graduated from the Politehnica University of Bucharest with a major in control engineering. Since 2004 he is a Professor at the Department of Automatic Control and Systems Engineering of the same University. He is the head of the Laboratory of Natural Computing and Robotics and of the Laboratory for Cognitive Robotics applied in Medicine. He holds a double doctorate in control engineering and in natural sciences, respectively. His research interests include robot control, multi-robot systems, human-robot interaction, membrane computing, and computational biology.

Index

Stay Current on the Latest Emerging Research Developments

Become an IGI Global Reviewer for Authored Book Projects

The overall success of an authored book project is dependent on quality and timely reviews.

In this competitive age of scholarly publishing, constructive and timely feedback significantly decreases the turnaround time of manuscripts from submission to acceptance, allowing the publication and discovery of progressive research at a much more expeditious rate. Several IGI Global authored book projects are currently seeking highly qualified experts in the field to fill vacancies on their respective editorial review boards:

Applications may be sent to:
development@igi-global.com

Applicants must have a doctorate (or an equivalent degree) as well as publishing and reviewing experience. Reviewers are asked to write reviews in a timely, collegial, and constructive manner. All reviewers will begin their role on an ad-hoc basis for a period of one year, and upon successful completion of this term can be considered for full editorial review board status, with the potential for a subsequent promotion to Associate Editor.

If you have a colleague that may be interested in this opportunity, we encourage you to share this information with them.

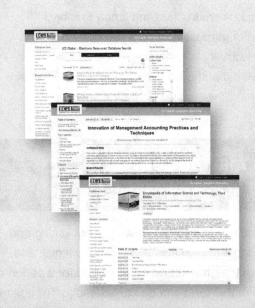

Printed in the United States
By Bookmasters